DAC Guidelines and Reference

Applying Strategic Environmental Assessment

GOOD PRACTICE GUIDANCE FOR DEVELOPMENT CO-OPERATION

OECD

ORGANISATION FOR ECONOMIC CO-OPERATION AND DEVELOPMENT

ORGANISATION FOR ECONOMIC CO-OPERATION AND DEVELOPMENT

The OECD is a unique forum where the governments of 30 democracies work together to address the economic, social and environmental challenges of globalisation. The OECD is also at the forefront of efforts to understand and to help governments respond to new developments and concerns, such as corporate governance, the information economy and the challenges of an ageing population. The Organisation provides a setting where governments can compare policy experiences, seek answers to common problems, identify good practice and work to co-ordinate domestic and international policies.

The OECD member countries are: Australia, Austria, Belgium, Canada, the Czech Republic, Denmark, Finland, France, Germany, Greece, Hungary, Iceland, Ireland, Italy, Japan, Korea, Luxembourg, Mexico, the Netherlands, New Zealand, Norway, Poland, Portugal, the Slovak Republic, Spain, Sweden, Switzerland, Turkey, the United Kingdom and the United States. The Commission of the European Communities takes part in the work of the OECD.

OECD Publishing disseminates widely the results of the Organisation's statistics gathering and research on economic, social and environmental issues, as well as the conventions, guidelines and standards agreed by its members.

This work is published on the responsibility of the Secretary-General of the OECD. The opinions expressed and arguments employed herein do not necessarily reflect the official views of the Organisation or of the governments of its member countries.

Also available in French under the title:
L'évaluation environnementale stratégique
GUIDE DE BONNES PRATIQUES DANS LE DOMAINE DE LA COOPÉRATION POUR LE DÉVELOPPEMENT

© OECD 2006

No reproduction, copy, transmission or translation of this publication may be made without written permission. Applications should be sent to OECD Publishing: *rights@oecd.org* or by fax (33-1) 45 24 99 30. Permission to photocopy a portion of this work should be addressed to the Centre français d'exploitation du droit de copie (CFC), 20, rue des Grands-Augustins, 75006 Paris, France, fax (33-1) 46 34 67 19, *contact@cfcopies.com* or (for US only) to Copyright Clearance Center (CCC), 222 Rosewood Drive Danvers, MA 01923, USA, fax (978) 646 8600, *info@copyright.com*.

Foreword

Sustainable development is a global issue that OECD and developing countries can only achieve through joint efforts. The OECD Development Assistance Committee (DAC) is a key forum where major bilateral and multilateral donors work together to increase the effectiveness of their common efforts to support sustainable development. The DAC focuses, in particular, on how international development co-operation can help developing countries combat poverty and participate in the global economy. The objective is to help shape the response of development co-operation at the policy and operational levels.

DAC work in the area of development and environment is carried out primarily through its Network on Environment and Development Co-operation (ENVIRONET). In 2004, ENVIRONET established a Task Team on Strategic Environmental Assessment (SEA), in response to the demand for guidance on the most efficient and effective application of SEA in the context of development co-operation. This Good Practice Guidance is the product of this work. It has involved a comprehensive consultation process with bilateral and multilateral development co-operation agencies, as well as representatives from partner countries and individual experts and practitioners from a wide range of developing and developed countries.

> **DAC members are:** Australia, Austria, Belgium, Canada, Denmark, Finland, France, Germany, Greece, Ireland, Italy, Japan, Luxembourg, Netherlands, New Zealand, Portugal, Norway, Spain, Sweden, Switzerland, United Kingdom, United States and Commission of the European Communities. The International Monetary Fund (IMF), the United Nations Development Programme (UNDP), and the World Bank participate in the work of the DAC as observers.

Acknowledgements

This Guidance has been produced by the collective efforts of the OECD DAC ENVIRONET SEA Task Team, and co-opted experts from both developed and developing countries.

Members of the SEA Task Team are: Jon Hobbs (DFID, London) (Chair); Linda Ghanime (UNDP, New York) (Vice-Chair); Harald Lossack, Jan-Peter Schemmel and Stephan Paulus (GTZ, Bonn and Berlin); Peter Croal, Bob Weir and Helene Gichenje (CIDA, Ottawa) and Tamara Levine (CIDA, Ottawa, now University of Sussex); Arnold Jacques de Dixmude (Ministry of Foreign Affairs, Belgium); Naïg Cozannet (AFD, Paris); Jouko Eskelinen and Matti Nummelin (Ministry of Foreign Affairs, Finland); Miriam Ciscar (AECS, Spain); Etienne Coyette, Simon Le-Grand and Francoise Villette (Environment Directorate, European Commission, Brussels); Ellen Shipley (DFAT, Australia); Kaoru Kanoyashi and Kojma Takeharu (Japan); Elspeth Tarp, Jan Riemer and Merete Pedersen (DANIDA, Copenhagen); Steve Bass (DFID, London, now IIED); Joseph Gamperl (KfW, Berlin); Inger-Marie Bjonness (Ministry of Foreign Affairs, Oslo, now Norwegian Delegation to the EU, Brussels) and Anne Kari Hansen Ovind (Ministry of Foreign Affairs, Oslo); Rob van den Boom (DGIS, Netherlands); Tomas Andersson (Sida); Daniel Slunge, Olaf Drakenberg and Anders Ekbom (University of Gothenburg/Sida Helpdesk); Kulsum Ahmed, David Hanrahan, Fernando Loayza and Jean Roger Mercier (World Bank, Washington DC); Laura Lee (UNDP, New York); James Leaton (WWF-UK); Alex Weaver (Council for Scientific and Industrial Research, South Africa); Elizabeth Brito (IADB, Washington DC); Paul Driver (consultant, UK); Jiri Dusik and Simona Kosikova (Regional Environment Centre for Central and Eastern Europe, Czech Republic); Hussein Abaza and Fulai Sheng (UNEP, Geneva); David Howlett (UNDP, Tanzania, now DFID East Kilbride); Peter Poulsen (DFID, London), Jean-Paul Penrose (DFID West Africa, now consultant in Mozambique), Richard McNally (DFID London, now consultant in Viet Nam) and Andrew McCoubray (DFID, Nairobi); Elizabeth Smith (EBRD, London); Gregory Woodsworth (UNDP, Nairobi, Kenya); Jean-Paul Ledant and Juan Palerm (EC-Environment Integration Help Desk in Development, Brussels); Rob Verheem (EIA Commission, Netherlands); Roger Gebbels and Tanya Burdett (UK ODPM, London); Alfred Eberhardt (consultant to GTZ, Germany); John Horberry (consultant, UK); Peter Nelson (Land Use Consultants, UK); Steve Smith (Scott Wilson Consultants, UK); Barry Sadler (consultant to UNEP, based in Canada); David Annandale (Murdoch University, Australia, now consultant to Government of Bhutan); Daniele Ponzi (Formerly AsDB, now with AfDB, Tunis); Dawn Montague (WWF USA); Bea Coolman (WWF International); Luc Hens (Free University of Brussels); Marianne Fernagut (GRID Arendal); Barry Dalal-Clayton (IIED, London); Peter Tarr (Southern African Institute for Environmental Assessment, Namibia).

Rémi Paris and Jenny Hedman of the OECD/DAC Secretariat provided substantive and managerial assistance to the Task Team while Maria Consolati provided invaluable secretarial assistance. Steve Bass and Pierre Giroux, former and current Chairs of ENVIRONET, provided guidance and advice throughout the exercise.

Technical secretariat services were provided by the International Institute for Environment and Development (IIED).

ACKNOWLEDGEMENTS

A considerable number of individuals, many from developing countries, participated in an international conference on SEA organised by the Task Team at Halong Bay, Viet Nam (2005), in workshops organised by the Task Team at the annual meetings of the International Association for Impact Assessment (IAIA) in Marrakech (2003) and Prague (2005), and in Task Team meetings held in Brussels, London, Paris and Stockholm.

The work of the Task Team has been made possible through financial support provided by The Royal Norwegian Ministry of Foreign Affairs, Belgian Ministry of Foreign Affairs, Finnish Ministry of Foreign Affairs, GTZ, Sida, UK DFID, UNDP, UNEP and the World Bank.

Table of Contents

Acronyms 12

Preface 14

Executive Summary 17

Part I

Chapter 1. **Introduction** 23

    1.1. New approaches to development co-operation 24
    1.2. SEA meets the challenge of more strategic development co-operation 24
    1.3. Legal requirements for SEA 25
    1.4. SEA supports donors' harmonisation and alignment efforts 26
    1.5. What does this Guidance seek to achieve? 27
    1.6. How should this Guidance be used? 27
    1.7. Agreed development approaches underpinning this Guidance 28

Chapter 2. **Understanding Strategic Environmental Assessment** 29

    2.1. Positioning SEA in the decision-making hierarchy 30
    2.2. How SEA has evolved to address the strategic levels of decision making 31
    2.3. SEA: A family of approaches using a variety of tools 32
    2.4. SEA: A continuum of applications 34
    2.5. The relationship of SEA with other policy appraisal approaches
        and supporting tools 35

    Notes 39

Chapter 3. **The Benefits of Using Strategic Environmental Assessment
in Development Co-operation** 41

    3.1. Supporting the integration of environment and development 42
    3.2. Identifying unexpected potential impacts of reform proposals 43
    3.3. Improving the identification of new opportunities 43
    3.4. Preventing costly mistakes 45
    3.5. Building public engagement in decision making for improved
        governance 45
    3.6. Facilitating transboundary co-operation 46
    3.7. Safeguarding environmental assets for sustainable development
        and poverty reduction 47

TABLE OF CONTENTS

Chapter 4. **Towards Strategic Environmental Assessment Good Practice: Principles and Processes** 49

    4.1. Basic principles for SEA. 50
    4.2. The institutional dimension of SEA 51
    4.3. Stages and steps for undertaking SEA at the plan and programme level 54

    Notes 61

Part II

Chapter 5. **Applications of Strategic Environmental Assessment in Development Co-operation** 65

    5.1. Key entry points for SEA 66
    5.2. Donor harmonisation for SEAs 69
    5.3. Guidance Notes 71

    A. Guidance Notes and Checklists for SEA Led by Partner Country Governments 72

Guidance Note and Checklist 1:	National Overarching Strategies, Programmes and Plans.	72
Guidance Note and Checklist 2:	National Policy Reforms and Budget Support Programmes	77
Guidance Note and Checklist 3:	National Sectoral Policies, Plans and Programmes	81
Guidance Note and Checklist 4:	Infrastructure Investments Plans and Programmes	86
Guidance Note and Checklist 5:	National and Sub-National Spatial Development Plans and Programmes	90
Guidance Note and Checklist 6:	Trans-National Plans and Programmes	93

    B. Guidance Notes and Checklists for SEA Undertaken in Relation to Donor Agencies' Own Processes 98

Guidance Note and Checklist 7:	Donors' Country Assistance Strategies and Plans.	98
Guidance Note and Checklist 8:	Donors' Partnership Agreements with other Agencies	101
Guidance Note and Checklist 9:	Donors' Sector-Specific Policies.	103
Guidance Note and Checklist 10:	Donor-Backed Public Private Infrastructure Support Facilities and Programmes	106

    C. Guidance Notes and Checklists for SEA in Other, Related Circumstances ... 111

Guidance Note and Checklist 11:	Independent Review Commissions (which have implications for donors' policies and engagement).	111
Guidance Note and Checklist 12:	Major Private Sector-Led Projects and Plans	115

Case example 5.1. Mainstreaming environment into Poverty Reduction Strategies – SEA of Poverty Reduction Strategy Papers: Uganda and Rwanda 75
Case example 5.2. Incorporating environmental considerations into Ghana's Poverty Reduction Strategy processes: SEA of Poverty Reduction Process 76
Case example 5.3. SEA of poverty reduction credit, Tanzania (budget support)....................................... 79
Case example 5.4. SEA for policy reform in the water and sanitation sectors in Colombia (development policy lending)................80
Case example 5.5. The Kenya Education Support Programme 83
Case example 5.6. Sector EA of Indonesia Water Sector Adjustment Loan (WATSAL)... 84
Case example 5.7. Energy Environment Review in Iran and Egypt 85
Case example 5.8. Regional environmental assessment of Argentina flood protection 88
Case example 5.9. The Sperrgebiet land use plan, Namibia 91
Case example 5.10. SEA of the Great Western Development Strategy, China 92
Case example 5.11. Transboundary environmental assessment of the Nile basin..................................... 95
Case example 5.12. Mekong River Commission Basin Development Plan......... 96
Case example 5.13. SEA in Sida's country strategy for Viet Nam 100
Case example 5.14. The DFID-WWF Partnership Programme Agreement......... 103
Case example 5.15. CIDA action plan for HIV/AIDS............................ 105
Case example 5.16. Environment due diligence for financial intermediaries, based on European Bank for Reconstruction and Development (EBRD) procedures 108
Case example 5.17. Environmental risk management of a community-driven development project – Programme National de Développement Participatif (PNDP) Cameroon 109
Case example 5.18. The World Commission on Dams 113
Case example 5.19. The Extractive Industries Review....................... 114
Case example 5.20. Nam Theun 2 Hydropower Project, Lao PDR 117
Case example 5.21. The potential of SEA in relation to major oil and gas investments 118

Notes 119

Part III

Chapter 6. How to Evaluate Strategic Environmental Assessment 123

6.1. The role of evaluation 124
6.2. Evaluating the delivery of envisaged outcomes 125
6.3. Evaluation as quality control check.................................. 126

Checklist 6.1. Key questions for evaluating the delivery of envisaged outcomes of a PPP 125
Checklist 6.2. Key questions for evaluation as a quality control check 126

Chapter 7. **Capacity Development for Strategic Environmental Assessment** 129

    7.1. Why is SEA capacity development needed?............................. 130
    7.2. Mechanisms for developing capacities for SEA in partner countries......... 131
    7.3. SEA capacity development in donor organisations...................... 138
    7.4. SEA as a foundation for capacity development and learning societies 138

    Case example 7.1. Capacity-building needs assessment for the UNECE SEA Protocol Implementation in five countries in the Eastern Europe, Caucasus and Central Asia region (EECCA) 131
    Case example 7.2. SEA training course in China 133
    Case example 7.3. SEA development in Mozambique...................... 134
    Case example 7.4. UNDP initiative for SEA capacity-building in Iran............ 134
    Case example 7.5. Assessing the potential to introduce SEA in Nepal.......... 135
    Case example 7.6. Results-based monitoring in the water and sanitation sector in Colombia.. 136
    Case example 7.7. Sofia Initiative on Strategic Environmental Assessment...... 137
    Case example 7.8. The SAIEA node model for EA support..................... 137
    Case example 7.9. Sida SEA Helpdesk – University of Gothenburg.............. 139
    Case example 7.10. Donor sharing of SEA experience 139

References and Bibliography.. 141

Annex A. **Glossary of Terms**... 145
Annex B. **Assessment Approaches Complementary to Strategic Environmental Assessment**... 149
Annex C. **Analytical and Decision-making Tools for Strategic Environmental Assessment**... 153
Annex D. **Selected Sources of Information on Strategic Environmental Assessment**... 159

List of boxes

1.1. The Millennium Development Goals and international trends: shape new ways of aid delivery 24
1.2. Harmonisation and alignment... 26
2.1. Defining policies, plans and programmes 31
2.2. Some examples of tools that could be used in SEA 33
2.3. Linkages between environmental stress and conflict 37
2.4. SEA in post-conflict countries 37
2.5. Country environmental analysis... 38
3.1. SEA benefits at a glance 42
3.2. Incorporating environmental considerations in Tanzania's second PRS 43
3.3. Achieving positive revision to forest policies in Ghana 44
3.4. Thermal Power Generation Policy, Pakistan: an early SEA would have helped..... 44
3.5. Sector EA of Mexican Tourism 45
3.6. SEA for Water Use, South Africa 46
4.1. Institutions centred SEA 53
4.2. Preparatory tasks in SEA... 55
7.1. Basic principles of capacity development 130

List of tables

2.1. SEA and EIA compared 32
4.1. Examples of policy reforms and potential environment linkages 59
5.1. Key entry points for SEA: Country level 67
5.2. Key entry points for SEA: Development agencies' own activities 67
5.3. SEA led by country governments and SEA undertaken in a donor agency's own processes: Key features compared 68
5.4. Key entry points for SEA: Related entry points 69
7.1. Capacity development framework for SEA. 133

List of figures

2.1. SEA: Up-streaming environmental considerations into the decision-making hierarchy 30
2.2. A continuum of SEA application 34
4.1. Steps to address institutional considerations in SEA 51
4.2. Basic stages in SEA. 54

Acronyms

CAS	Country assistance strategy
CCA	Common country assessment
CEA	Country environmental analysis
CEA	Cumulative effects assessment
CEE	Central and Eastern Europe
CIA	Cumulative Impacts Assessment
CSO	Civil society organisation
DAC	Development Assistance Committee (OECD)
DoC	Drivers of change
DBS	Direct budget support
DPL	Development policy lending
EA	Environmental assessment
EC	European Commission
EIA	Environmental impact assessment
EMP	Environmental management plan
ESIA	Environmental and social impact assessment
EU	European Union
GIS	Geographic information system
HIA	Health impact assessment
IAIA	International Association for Impact Assessment
JAS	Joint Assistance Strategies
JPOI	Johannesburg Plan of Implementation
MDG	Millennium Development Goal
NGO	Non-governmental organisation
NSDS	National sustainable development strategy
OECD	Organisation for Economic Co-operation and Development
OP	Operational policy (World Bank)
PCIA	Peace and Conflict Assessment
PIA	Poverty impact assessment
PPP	Policy, plan and programme. (In this publication, PPP is used in an expanded sense to include strategies and [where appropriate] macro projects.)
PRS	Poverty reduction strategy
PRSC	Poverty reduction strategy credit
PRSP	Poverty reduction strategy paper
PSIA	Poverty and social impact analysis
REA	Regional environmental assessment
SCA	Strategic conflict assessment
SEA	Strategic environmental assessment

SIA	Social impact assessment
SIA	Strategic impact assessment
SIA	Sustainability impact assessment
SWAp	Sector-wide approach
TEA	Transboundary environmental assessment
TOR	Terms of reference
UNDAF	United Nations Development Assistance Framework
UNECE	United Nations Economic Commission for Europe
WSSD	World Summit on Sustainable Development

Preface

Poor people in developing countries often depend more directly on natural resources than any other group in society. They are usually the first to suffer when those resources are damaged or become scarce. This means it is vital that we consider the environment in all our development work. How we manage the environment will affect the long-term success of development and play a significant part in our progress towards the Millennium Development Goals (MDGs).

This is why MDG 7 commits us to ensuring environmental sustainability. It demands that we make the principles of sustainable development an integral part of our policies and programmes. We must consider the environment when making decisions, just as we consider economic and social issues. Strategic Environmental Assessment (SEA) is the most promising way to make this happen.

SEA helps decision makers reach a better understanding of how environmental, social and economic considerations fit together. Without that understanding, we risk turning today's development successes into tomorrow's environmental challenges. In short, SEA helps decision makers think through the consequences of their actions.

In March 2005, ministers and heads of development agencies from more than 100 developed and developing countries met in Paris to consider ways to make aid work better. As a result, we adopted the Paris Declaration on Aid Effectiveness.

The Declaration sets out our plan for improving the way we deliver and manage our co-operation on development. It says that aid needs to be more effective and its delivery better coordinated. It calls for more and better support for developing countries' priorities. It underlines the importance of working through existing institutions in developing countries whenever possible, to build their own development capacity, instead of creating parallel paths.

The Paris Declaration also commits donors and their partner countries to "develop and apply common approaches for Strategic Environmental Assessment". This good practice guidance on applying SEA in development co-operation is the first step along this road. We are pleased that our agencies have risen to this challenge so promptly and joined forces with a wide range of partners to turn this commitment into action.

The International Association for Impact Assessment (IAIA) has given the OECD/DAC Environet SEA Task Team its prestigious Institutional Award for 2006. The IAIA citation acknowledges the advances this work has made in environmental management and commends the outstanding team work that made it possible.

We take special pride in the role of our agencies in leading this process. But we also know this document is the result of sustained efforts by many participating agencies and individuals, to whom we offer our sincere thanks.

We encourage all decision-makers involved in development co-operation work to apply this guidance. It is a practical way to help improve development effectiveness and achieve lasting results for poor people.

Rt. Hon. Hilary Benn MP
Secretary of State
for International Development
London
United Kingdom

Kemal Derviş
Administrator
UNDP
New York

Richard Manning
Chair of the OECD
Development Assistance
Committee (DAC)
Paris
France

ISBN 92-64-02657-6
Applying Strategic Environmental Assessment
Good Practice Guidance for Development Co-operation
© OECD 2006

Executive Summary

1. Introduction

Development assistance is increasingly being provided through strategic-level interventions, aimed to make aid more effective. To ensure environmental considerations are taken into account in this new aid context, established environmental assessment tools at the project level need to be complemented by approaches fully adapted to policies, plans and programmes. Strategic Environmental Assessment (SEA) meets this need.

SEA provides a practical and direct means of progressing MDG 7 on Environmental Sustainability (agreed at the UN General Assembly in 2000). This calls for the "integration of the principles of sustainable development into country policies and programmes". Secondly, SEA also helps further the Johannesburg Plan of Implementation agreed at the World Summit on Sustainable Development in 2002, which stressed the importance of "strategic frameworks and balanced decision making […] for advancing the sustainable development agenda".

The Paris Declaration on Aid Effectiveness, adopted in 2005, commits donors to reform the way in which aid is delivered to improve effectiveness, by harmonising their efforts and aligning behind partner countries' priorities. It also calls upon donors and partners to work together to "*develop and apply common approaches for strategic environmental assessment at sector and national levels*".

This Guidance aims to respond to these challenges. Drawing on practical experience and established "good practice", it points to ways to support the application of SEA in the formulation and assessment of development policies, plans and programmes. In view of the great diversity of circumstances across different countries, it seeks to provide a commonly-agreed and shared model that allows for flexibility in developing appropriate applications of SEA to the diversity of needs. It is presented in the context of a rapidly emerging framework of international and national legislation on SEA in both developed and developing countries.

2. Understanding SEA

SEA refers to a range of "analytical and participatory approaches that aim to integrate environmental considerations into policies, plans and programmes and evaluate the inter linkages with economic and social considerations". SEA can be described as a family of approaches which use a variety of tools, rather than a single, fixed and prescriptive approach. A good SEA is adapted and tailor-made to the context in which it is applied. This can be thought as a continuum of increasing integration: at one end of the continuum, the principle aim is to integrate environment, alongside economic and social concerns, into strategic decision making; at the other end, the emphasis is on the full integration of the environmental, social and economic factors into a holistic sustainability assessment.

EXECUTIVE SUMMARY

SEA is applied at the very earliest stages of decision making both to help formulate policies, plans and programmes and to assess their potential development effectiveness and sustainability. This distinguishes SEA from more traditional environmental assessment tools, such as Environmental Impact Assessment (EIA), which have a proven track record in addressing the environmental threats and opportunities of specific projects but are less easily applied to policies, plans and programmes. SEA is not a substitute for, but complements, EIA and other assessment approaches and tools.

3. The benefits of using SEA

Applying SEA to development co-operation has benefits for both decision-making procedures and development outcomes. It provides the environmental evidence to support more informed decision making, and to identify new opportunities by encouraging a systematic and thorough examination of development options. SEA helps to ensure that the prudent management of natural resources and the environment provide the foundations for sustainable economic growth which, in turn, support political stability. SEA can also assist in building stakeholder engagement for improved governance, facilitate trans-boundary co-operation around shared environmental resources, and contribute to conflict prevention.

4. Towards good practice in SEA

SEA is a continuous, iterative and adaptive process focussed on strengthening institutions and governance. It is not a separate system, nor a simple linear, technical approach. Instead, it adds value to existing country systems and reinforces their effectiveness by assessing and building capacity for institutions and environmental management systems.

Where SEA is applied to plans and programmes, a structured approach to integrating environmental considerations can be used. Key stages for carrying out an SEA on the level of plans or programmes include: establishing the context, undertaking the needed analysis with appropriate stakeholders, informing and influencing decision making, and monitoring and evaluation. SEA applied at the policy level requires a particular focus on the political, institutional and governance context underlying decision-making processes.

5. Application of SEA in development co-operation

The shift of emphasis away from development projects to programme and policy support has created a number of particular entry points for the application of SEA. This guidance outlines the benefits of using SEA in a range of different circumstances, and sets out 12 key "entry points" for effective application of SEA to development co-operation. It points to key questions to be addressed for each of them, accompanied by specific checklists of these questions, and illustrative case examples.

The entry points for SEA can be grouped into:

1. *Strategic planning processes led by a developing country:* These include national overarching strategies, programmes and plans; national policy reforms and budget support programmes; sectoral policies, plans and programmes; infrastructure investments plans and programmes; national and sub-national spatial development plans and programmes and transnational plans and programmes.

2. *Development agencies' own processes:* These include donors' country assistance strategies and plans; partnership agreements with other donor agencies, donors' sector-specific policies, and donor-supported public-private infrastructure support facilities and programmes.

3. *Other related circumstances:* These include independent Review Commissions and major private sector-led projects and plans.

6. How to evaluate an SEA

The key deliverable of an SEA is a process with development outcomes, not a product. Quality control therefore considers how well procedures have been carried out. But in the long term, the achievement of development outcomes, while ensuring the maintenance of environmental sustainability, will be the key measure of success.

When reviewing SEA processes, key questions concern: the quality of information, level of stakeholder participation, defined objectives of the SEA, assessment of environmental impacts, planned follow-up activities, and constraints.

Key questions to help evaluators focus on development outcomes of an SEA relate to: the accuracy of assumptions made during the SEA; its influence on the PPP process, on the implementation process, on development goals and on accountability; and the outcome of capacity-building activities.

7. Developing the capacity for effective use of SEA

Experiences of applying SEA have repeatedly highlighted two key challenges: lack of awareness of the value and importance of SEA, and, when the value is appreciated, lack of knowledge on how to implement SEA. These challenges can be significantly addressed by capacity development for SEA in both development agencies and partner countries.

For capacity development in partner countries, a capacity needs assessment is the first step. Support involves activities such as technical training, awareness-raising workshops, supporting the institutionalisation of the SEA process and its evaluation systems, and networking for sharing experiences.

Capacity development in donor organisations may go through training activities for staff and SEA guidelines and support, as well as systematic reviews and evaluations.

Quick reference guide

What is SEA and why is SEA relevant to the international development agenda?	Chapters 1 and 2
What are the potential benefits of using SEA?	Chapter 3
What are the basic principles and processes involved in SEA?	Chapter 4
Where can SEA be effectively applied?	Chapter 5
What constitutes a good SEA process?	Chapter 6
How can we develop the capacity to apply SEA?	Chapter 7
Where is more information available?	Annexes and at *www.seataskteam.net*

Part I

Chapter 1.	**Introduction**	23
Chapter 2.	**Understanding Strategic Environmental Assessment**	29
Chapter 3.	**The Benefits of Using Strategic Environmental Assessment in Development Co-operation**	41
Chapter 4.	**Towards Strategic Environmental Assessment Good Practice: Principles and Processes**	49

ISBN 92-64-02657-6
Applying Strategic Environmental Assessment
Good Practice Guidance for Development Co-operation
© OECD 2006

PART I

Chapter 1

Introduction

1.1. New approaches to development co-operation

The way international development assistance is provided is changing. The aim is to make aid more effective in supporting progress towards the Millennium Development Goals (MDGs) and to meet the needs of the poor. This involves a shift towards strategic interventions, in line with the Johannesburg Plan of Implementation (see Box 1.1). Increasingly, development co-operation agencies provide support at the level of policies, plans and programmes (PPPs). This includes, in particular, supporting comprehensive development frameworks such as "poverty reduction strategies" which are formulated and led by the developing partner country and implemented through national and local systems and institutions.

> **Box 1.1. The Millennium Development Goals and international trends: shape new ways of aid delivery**
>
> Current international efforts to reduce global poverty focus on the Millennium Development Goals (MDGs), endorsed by the UN General Assembly in 2000. A number of the MDGs provide an impetus for a strategic approach to environmental sustainability.* In particular, MDG 7 on environmental sustainability recognises the need to:
>
> "… *integrate the principles of sustainable development into country policies and programmes and reverse the loss of environmental resources.*"
>
> The Johannesburg Plan of Implementation agreed at the World Summit on Sustainable Development in 2002, stresses "the importance of strategic frameworks and balanced decision making as fundamental requirements for advancing the sustainable development agenda".
>
> * For a comprehensive review of the links between MDGs and Strategic Environmental Assessment refer to IIED (2004), particularly Chapter 1.

1.2. SEA meets the challenge of more strategic development co-operation

Development agencies have years of experience in using environmental impact assessment (EIA) to integrate environmental concerns into the projects which they support. As compared to individual projects, however, strategic-level interventions, notably at the policy-level, are much more influenced by political factors than by technical criteria. Moreover, the environmental effects associated with policy reforms are often indirect, occur gradually over the long term and are difficult to assess accurately. While still very valuable and relevant at the project level, established EIA procedures, methods and techniques have only limited application at the level of policies, plans and programmes.

For these and other reasons the shift towards new development co-operation instruments such as direct budgetary support, policy reform, and sector-wide support programmes has created a need for different environmental assessment approaches. Strategic Environmental Assessment (SEA) – a range of "*analytical and participatory approaches that aim to integrate environmental considerations into policies, plans and programmes and evaluate the inter linkages with economic and social considerations"* – responds to this need. It allows the integration of environmental considerations – alongside social and economic aspects – into strategic decision making at all stages and tiers of development co-operation. SEA is not a substitute for traditional project impact assessment tools, but a complement to them.

Chapter 2 below explains how SEA can support the integration of environmental, social and economic factors into strategic decision making processes. Chapter 3 provides examples of how SEA has improved the process of decision-taking, making important contributions to development effectiveness.

1.3. Legal requirements for SEA

This Guidance is presented in the context of an emerging framework of international and national legislation on SEA in both developed and developing countries. Two important international instruments now prescribe the application of SEA. Firstly, the European Directive (2001/42/EC) on the *Assessment of the Effects of Certain Plans and Programmes on the Environment*, known as the SEA Directive, came into effect in 2004 and applies to all 25 member states of the European Union. It requires an environmental assessment for certain plans and programmes at various levels (national, regional and local) that are likely to have significant effects on the environment. Secondly, a similar provision is contained in the SEA Protocol to the Espoo Convention (UNECE Convention on EIA in a Transboundary Context), agreed in Kiev in May 2003. The Protocol includes a separate article encouraging the use of SEA in the context of policies and legislation. It will become effective once ratified by at least 16 countries.

Many developed and developing countries have either national legislative or other provisions for SEA, *e.g.* statutory instruments, cabinet and ministerial decisions, circulars and advice notes. A number of EU countries had such provisions even prior to the above-mentioned SEA Directive taking effect. Several non-EU European countries also have legal requirements to apply SEA. The EU accession process for candidate countries, as well as ratification of the SEA Protocol to the Espoo Convention, is likely to make comparable legal requirements more widespread. In Canada, there is an administrative requirement to conduct SEA on all PPPs through a Cabinet Level Directive. In the USA, programmatic environmental assessment is required for large projects and programmes.

Increasingly, developing countries are introducing legislation or regulations to undertake SEA – sometimes in EIA laws and sometimes in natural resource or sectors laws and regulations. In South Africa, some sectoral and planning regulations identify SEA as an approach for integrated environmental management. In the Dominican Republic, legislation refers to SEA or strategic environmental evaluation. Elsewhere, the existing EIA legislation requires an SEA-type approach to be applied either to plans (*e.g.* China), programmes (*e.g.* Belize) or both policies and programmes (*e.g.* Ethiopia). The Convention on Biological Diversity (Article 6b and Article 14) encourages the use of SEA in its implementation without making it a specific requirement.*

Development co-operation agencies need to take account of these trends.

---

* See also Decision on SEA at COP 8 of the CBD *www.biodiv.org/decisions*.

1.4. SEA supports donors' harmonisation and alignment efforts

As SEA is becoming more widely adopted by donor agencies and their developing country partners, the donor community is committed to harmonising its procedures and requirements in this area. The *Paris Declaration on Aid Effectiveness*, adopted on 2 March 2005, commits the donor community to reforming the ways in which aid is delivered and to working in closer harmony to enhance development efficiency and effectiveness.

The Paris Declaration calls upon development agencies and partner countries to develop common approaches to environmental assessment generally, and to SEA specifically (*www.oecd.org/dac*):

"Donors have achieved considerable progress in harmonisation around environmental impact assessment (EIA) including relevant health and social issues at the project level. This progress needs to be deepened, including on addressing implications of global environmental issues such as climate change, desertification and loss of biodiversity.

Development agencies and partner countries jointly commit to:

- *Strengthen the application of EIAs and deepen common procedures for projects, including consultations with stakeholders; and develop and apply common approaches for "strategic environmental assessment" at the sector and national levels.*

- *Continue to develop the specialised technical and policy capacity necessary for environmental analysis and for enforcement of legislation."*

This Guidance responds to this challenge by providing a framework for greater consensus in the development and application of SEA and coherence with related and complementary policy appraisal tools and procedures.

The Paris Declaration also emphasises the need for donor agencies to better align behind the priorities of developing countries and their strategies to address these priorities. This Guidance supports this objective by pointing to SEA approaches specifically tailored to the types of development PPPs applied by developing partner countries. It provides advice on how to analyse the potential environmental risks and benefits of such PPPs and ensure the engagement of relevant stakeholders.

Box 1.2. Harmonisation and alignment

Alignment and *Harmonisation* are complementary and mutually supportive processes being pursued in line with the commitments made in the Paris Declaration on Aid Effectiveness. But they are different. For *alignment*, the focus is on making use of partners' own procedures. For *harmonisation*, the focus is on development agencies getting together to develop similar or common procedures.

In the area of SEA, these differences do not have much practical impact: SEA is an emerging field and no firm procedures exist either for development agencies or partners. Moreover, as this Guidance stresses, there should be no single "blueprint" procedure for SEA, as may exist in fields such as financial management or accounting. There is an opportunity for development agencies and partners to jointly define key principles underlying SEA that will lead to approaches applicable to all countries. This Guidance seeks to translate this opportunity into reality.

1.5. What does this Guidance seek to achieve?

The different needs of SEA users, the different legal requirements they face, the diversity of applications of SEA in development co-operation, and, last but not least, the rapid evolution of SEA practice imply that it is neither feasible nor desirable to suggest a precise "one size fits all" methodology let alone prescriptive, blue print guidelines for SEA.

Accordingly, this Guidance provides a flexible framework around common-agreed principles to help bring coherence to SEA practice. Chapter 4 below outlines guiding principles and key generic steps for SEA. These reflect emerging consensus on the nature, role and application of SEA in the context of development co-operation. Drawing on current international experience, it aims to promote and support the practical use of SEA in the formulation and assessment of development PPPs – whether by the donor agencies or their partner governments. Chapter 5 below sets out 12 key potential "entry points" for SEA application).

This Guidance aims to:

- Describe the importance of environmental considerations in underpinning sustained economic growth and poverty reduction.
- Explain the environment's contribution to the sustainability of a range of development interventions.
- Describe the emerging consensus about the value of SEA to development co-operation – drawing on evolving international experience – including generally agreed key principles and procedures.
- Identify key "entry points" for effective application of SEA to development co-operation and key questions to be addressed for each of them.
- Illustrate good practice through illustrative cases.
- Identify institutional capacity needs and opportunities to respond through capacity development.
- Provide sources of further information.

The ultimate objectives are to ensure that:

- Environmental considerations, and their linkages with social and economic factors, are adequately understood, recognising the contribution of environmental management to economic growth and poverty reduction.
- Environmental and social considerations are appropriately analysed and taken into account in development policy, planning and strategic decision making at the formative stage and appropriate response measures, effectively integrated into the development of PPPs and projects.
- As a result of the above, the outcomes of PPPs have better prospects to contribute to sustainable development and attainment of the MDGs. (Chapter iii provides insights as to how to assess the effectiveness of SEA in his regard.)

Special emphasis has also been placed on partner countries' capacity development needs (see Chapter iii).

1.6. How should this Guidance be used?

Given the current status of SEA and the dynamics of international development, this Guidance should be regarded as a living' document to be updated as significant new experiences emerge. It is intended as a point of reference rather than a prescriptive blueprint for all development agencies, in all countries, at all times. The recommended approaches should be considered on a case-by-case basis, and adapted to reflect partner country circumstances, development agencies' mandates, and the specificities of the PPPs being considered. On-going developments may be found on the DAC Task Team's web site (*www.seataskteam.net*).

This Guidance is aimed primarily at professionals working in development agencies and developing country government departments directly involved in PPP development and assessment, but it will also be of value more generally to other policy analysts and planners.

1.7. Agreed development approaches underpinning this Guidance

Three overarching principles underpin this Guidance and indeed all development co-operation efforts:

- ***Partner country ownership:*** each country is responsible for its own development, and the donors' role is to *support, not substitute,* national efforts.
- ***Development agencies need to act within an explicit, government-owned strategic framework and programme:*** fragmented and piecemeal initiatives by development agencies are unlikely to be effective. This requires that development agencies work jointly with governments *and* with each other.
- ***Development agencies need to be sensitive to country context:*** taking account of country-specific institutions, priorities, and national legal requirements or international commitments for SEA. Agencies should recognise that donor behaviour may affect the in-country accountability relations of governments, the relations between tiers of government, and the relations between various government agencies. Effective institutional development, in particular, should start from the premise of building on what already exists rather than transplanting entirely new systems. Good practices continue to evolve and need to be adapted to the circumstances of specific agencies and countries.

Development agencies need to tailor their internal policies and procedures to be consistent with these requirements.

ISBN 92-64-02657-6
Applying Strategic Environmental Assessment
Good Practice Guidance for Development Co-operation
© OECD 2006

PART I

Chapter 2

Understanding Strategic Environmental Assessment

I.2. UNDERSTANDING STRATEGIC ENVIRONMENTAL ASSESSMENT

This Guidance uses the term SEA to describe analytical and participatory approaches that aim to integrate environmental considerations into policies, plans and programmes and evaluate the inter linkages with economic and social considerations.

This chapter provides a model to describe these approaches and differentiate them from related, complementary approaches to environmental and social assessment. It also explains how SEA can be applied across a *continuum* of increasing integration (see Figure 2.2), whether the principal aim is to integrate environment into strategic decision making or the full integration of environmental, social and economic factors in more holistic sustainability appraisal.

The practice of SEA is largely shaped by the diverse circumstances in which it is applied and by the demands it addresses. No matter how SEA is being used, it has some universal principles (see Section 4.1).

2.1. Positioning SEA in the decision-making hierarchy

There is a hierarchy of levels in decision making comprising projects, programmes, plans and policies (see Figure 2.1). Logically, policies shape the subsequent plans, programmes and projects that put those policies into practice. Policies are at the top of the decision-making hierarchy. As one moves down the hierarchy from policies to projects, the nature of decision-making changes, as does the nature of environmental assessment needed. Policy-level assessment tends to deal with more flexible proposals and a wider range of scenarios. Project-level assessment usually has well defined and prescribed specifications.

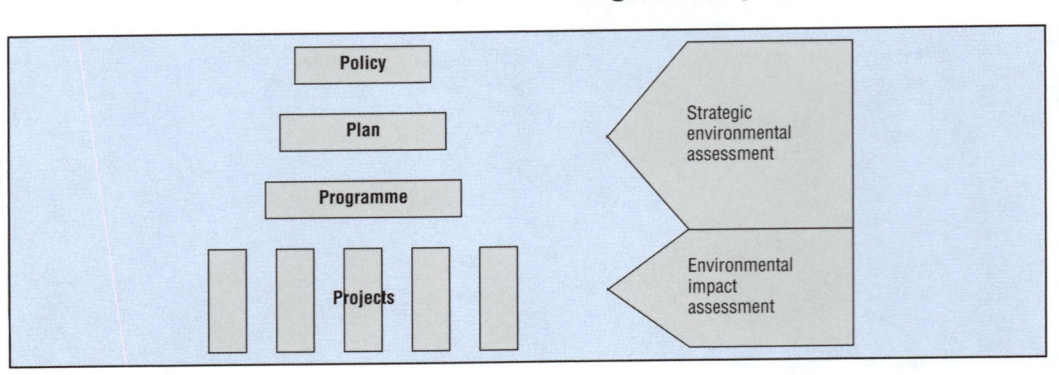

Figure 2.1. **SEA: Up-streaming environmental considerations into the decision-making hierarchy**

Policies, plans and programmes (PPPs) (see Box 2.1) are more "strategic" as they determine the general direction or approach to be followed towards broad goals. SEA is

> **Box 2.1. Defining policies, plans and programmes**
>
> **Policy.** A general course of action or proposed overall direction that a government is or will be pursuing and that guides ongoing decision making.
>
> **Plan.** A purposeful forward looking strategy or design, often with co-ordinated priorities, options and measures that elaborate and implement policy.
>
> **Programme.** A coherent, organised agenda or schedule of commitments, proposals, instruments and/or activities that elaborate and implement policy.
>
> *Source:* Sadler and Verheem (1996).

applied to these more strategic levels. Environmental Impact Assessment (EIA) is used on projects that put PPPs into tangible effect.

2.2. How SEA has evolved to address the strategic levels of decision making

EIA has been a proven tool over the past 35 years and has a good track record in evaluating the environmental risks and opportunities of project proposals and improving the quality of outcomes. Yet the need for a similar assessment process at the strategic level of decision-making has been recognised. Leaving environmental assessment until the project stage severely limits the opportunities to identify the strategic choices that might lead to more sustainable outcomes and reduce risks to the environmental resource base that provides the foundations for growth and development. While EIA has been successful in integrating environmental considerations in development projects, EIA practices still pose a challenge for many developing countries.

Project assessment invariably takes place in a predetermined policy environment. For example, an EIA of a new fossil fuel energy generation plant will be unlikely to consider other energy generating possibilities. Project EIAs are usually prepared by the project proponent with a vested interest in project approval. So, alternatives for energy generation will be limited to location and technology choices within the framework of fossil fuel generation. A strategic decision will already have been made to develop fossil fuel energy resources as the preferred option, and this will likely have been taken without rigorous attention to environmental considerations.

At best, an EIA will describe a "no or without project" option. This will often serve only to motivate the proposal rather than provide adequate consideration of the full range of options. The main elements of the project are taken as a "given". By addressing choices upstream of projects in the realm of PPPs, SEA is able to consider a far richer array of development options. In this way, SEA directly influences the policy environment, preferably in its formative stages, and increases the likelihood of creating sustainable development outcomes and reduced environmental risks.

The practical application of EIA has led to the emergence of two spin-off assessment approaches: Social Impact Assessment and Cumulative Impact Assessment:

- **Social Impact Assessment (SIA).** The emphasis of early EIAs was on biophysical concerns (in more recent years, EIA practice has interpreted the environment' more holistically). As a result, SIA emerged as a technique to give explicit attention to the social dimensions in project assessment. It has been used either as a stand-alone approach or in a more integrated application as Environmental and Social Impact

Assessment. This is particularly important in development co-operation where the interrelationships between poverty/development and environment have been well articulated.

- **Cumulative Impact Assessment (CIA).** EIA of a specific project proposal may fail to consider its potential aggregate, incremental and synergistic impacts with other projects in an area-wide programme of developments. CIA is a developing sub-set of SEA that has evolved as a way to capture these wider implications in project assessment. These implications may change the conclusions of an assessment of an individual project. A project considered to have few or insignificant impacts when considered on its own may, instead, be judged to have potential impacts of great significance when viewed as a part of a more complex set of developments. CIA therefore represents a move up the decision-making hierarchy, albeit to the programme/plan level, not the policy level.

It is important to note that SEA is not a substitute for EIA, SIA or CIA, but complements them. They are all integral parts of a comprehensive environmental assessment tool box. This has important implications in developing countries where EIA and SIA systems may still be in the process of being established. SEA application should not distract or complicate this process. Developing countries are frequently constrained by lack of resources to carry out project EIAs. SEA can, in fact, help to speed up EIA procedures and streamline their scope (and costs) by ensuring that project proposals are set within a policy framework that has already been subject to environmental scrutiny. This higher-level assessment process can consider and agree the most conducive strategy to enhance developmental outcomes and reduce negative impacts.

Table 2.1 compares and contrasts SEA and EIA and summarises their roles in decision making.

Table 2.1. **SEA and EIA compared**

EIA	SEA
Applied to specific and relatively short-term (life-cycle) projects and their specifications.	Applied to policies, plans and programmes with a broad and long-term strategic perspective.
Takes place at early stage of project planning once parameters are set.	Ideally, takes place at an early stage in strategic planning.
Considers limited range of project alternatives.	Considers a broad range of alternative scenarios.
Usually prepared and/or funded by the project proponents.	Conducted independently of any specific project proponent.
Focus on obtaining project permission, and rarely with feedback to policy, plan or programme consideration.	Focus on decision on policy, plan and programme implications for future lower-level decisions.
Well-defined, linear process with clear beginning and end (*e.g.* from feasibility to project approval).	Multi-stage, iterative process with feedback loops.
Preparation of an EIA document with prescribed format and contents is usually mandatory. This document provides a baseline reference for monitoring.	May not be formally documented.
Emphasis on mitigating environmental and social impacts of a specific project, but with identification of some project opportunities, off-sets, etc.	Emphasis on meeting balanced environmental, social and economic objectives in policies, plans and programmes. Includes identifying macro-level development outcomes.
Limited review of cumulative impacts, often limited to phases of a specific project. Does not cover regional-scale developments or multiple projects.	Inherently incorporates consideration of cumulative impacts.

2.3. SEA: A family of approaches using a variety of tools

Legal, procedural, institutional and political factors in different circumstances and countries will generally determine the way in which SEA is defined and applied. The availability of data, level of definition of the PPP, knowledge of direct and indirect impacts,

I.2. UNDERSTANDING STRATEGIC ENVIRONMENTAL ASSESSMENT

> **Box 2.2. Some examples of tools that could be used in SEA**
>
> - Tools for ensuring full stakeholder engagement:
>   - Stakeholder analysis to identify those affected and involved in the PPP decision
>   - Consultation surveys
>   - Consensus building processes
> - Tools for predicting environmental and socio-economic effects:
>   - Modelling or forecasting of direct environmental effects
>   - Matrices and network analysis
>   - Participatory or consultative techniques
>   - Geographical information systems as a tool to analyse, organise and present information
> - Tools for analysing and comparing options:
>   - Scenario analysis and multi-criteria analysis
>   - Risk analysis or assessment
>   - Cost benefit analysis
>   - Opinion surveys to identify priorities
>
> *These and a selection of other decision-making tools are described in* Annex C.

and available time frame for the SEA, will also help determine the approach taken and the tools used (see Box 2.2).

As a consequence, SEA can be applied in various ways to suit particular needs, *e.g.* some SEAs:

- Are "stand-alone" processes running *parallel* to core planning processes, while others are *integrated* into the planning, policy/decision-making processes.
- May focus on environmental impacts while others integrate all three dimensions of sustainability: environment, social and economic.
- May be applied to *evaluate an existing PPP*, or one that is about to be revised, to ascertain its environmental consequences. May *provide inputs into developing a PPP* – supporting, facilitating and improving its development (or revision) so that it addresses environmental dimensions effectively.
- May engage a broad range of stakeholders or be limited to expert policy analysts.
- Can be conducted in a short time frame or over a long period.
- May consist of a quick analysis while others require detailed analysis.
- Can be a finite, output-based activity (*e.g.* a report), or a more continuous process that is integrated within decision making, focused on outcomes, and that strengthens institutional capacity.

Moreover, different institutions use their own terminology to interpret SEA. Frequently, SEA approaches are given different, institution-specific "labels" such as sustainability appraisal, integrated assessment, strategic impact assessment, etc.

With such a wide variety of definitions and interpretations of SEA, and given the need to select an SEA approach and associated tools to suit the particular decision-making context, it would be inappropriate to suggest a rigorous and universally applicable definition of SEA. The flexible definition of SEA, highlighted at the start of this Chapter, is consistent with the full range of SEA approaches being applied in practice, and provides an umbrella term for a family of approaches. More precise definition is given to SEA by the principles and procedural criteria it employs (see Chapter 4) and by the nature of their application (see Chapter 5).

Similarly, there is no recipe approach to SEA. The diversity of applications reflects the need to adapt the concept to the need being addressed and the circumstances in which SEA is being applied. For all these reasons this document provides *guidance* rather than detailed *guidelines* on how to conduct an SEA.

2.4. SEA: A continuum of applications

To accommodate different approaches SEA is applied at various points on a **continuum** (Figure 2.2).

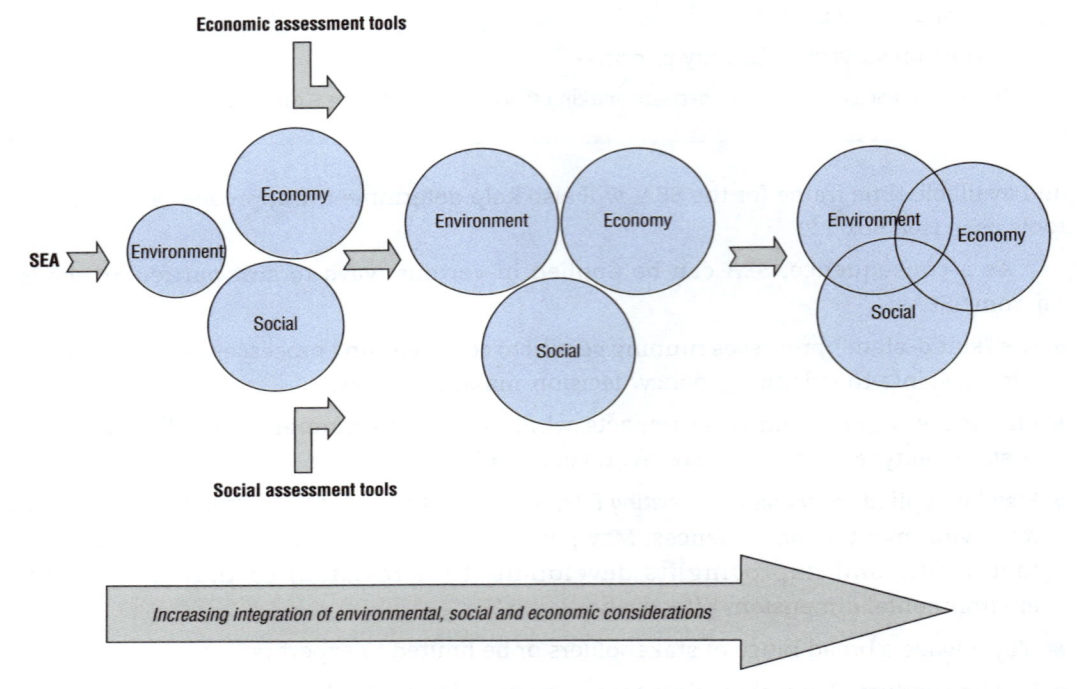

Figure 2.2. **A continuum of SEA application**

Notes:
1. The increasing circle size implies the "weight" given to environment. The overlapping indicates the extent of integration.
2. The right hand end of the continuum implies true sustainability where all three pillars of sustainability are given equal "weight" and are fully integrated.
3. The aim of environmental mainstreaming has first been, in the first instance, to get "environmental" considerations addressed in policy-making, planning and decision-taking; and then to promote increasing integration in addressing environmental, social and economic considerations.
4. Progress is being made in the application of key environmental, social and economic strategic assessment tools towards increasing integration.

At one end of the continuum the focus is mainly on **environmental integration**. It is characterized by the priority goal of mainstreaming and up-streaming environmental considerations into strategic decision making at the earliest stages of planning processes. This would be the priority when developing or evaluating PPPs that are primarily social and/or economic, but whose success would be considerably impaired without consideration of environmental constraints and opportunities. The first generation of Poverty Reduction Strategies provide an example of this. The usual approach adopted is one of a stand-alone SEA.

At the other end of the continuum, the focus is on an **integrated assessment** of the environmental, social and economic factors, and sometimes even broader factors such as institutions and governance dimensions. This is particularly relevant in developing countries, where environment takes on a meaning beyond the biophysical aspects to those more closely linked with quality of life and growth. This approach is sometimes called **sustainability appraisal/assessment**. In this situation, it is more usual for SEA principles (see Section 4.1) to be incorporated into policy analysis (examining a government's own administrative, decision-making and planning processes) rather than applied through a separate process.

The two approaches differ only in emphasis, but they imply a phased progression towards increasing complexity, balance and integration. Current experience is dominated by the environmental integration approach. Application of a more holistic, integrated approach is often constrained by institutional barriers.

The use of the continuum model does not imply that one specific approach is superior or inferior, or that they are totally distinct. The most appropriate approach to SEA will depend on the particular circumstances to be faced. Hence the position of the SEA approach adopted along the continuum will vary. The worst outcome would be failure to integrate environment into strategic decision making or to consider the linkages between the "pillars" of sustainability, and thus risk policy failure, missed opportunities, wasted resources and even unanticipated negative outcomes.

In addition to using the continuum model to describe the progression to integrated assessment, stress is increasingly being placed on the need for SEA to be more "institution-centred" or "decision-oriented" and to focus on the earliest stages of PPP formulation. This is particularly important in the case of Policy-level SEA (see Box 4.1).

Given the growing emphasis on development outcomes and results, it is increasingly important to develop institutional capacity over time to implement the participatory and analytical processes encapsulated in SEA to influence better decision making. There are a growing number of examples of the institutionalisation of SEA processes. In Ghana, for example an SEA of the Poverty Reduction Strategy evolved from "output-based" SEA to a continuous process SEA (see *Guidance Note 1* in Chapter 5).

2.5. The relationship of SEA with other policy appraisal approaches and supporting tools

There is a need for SEA to recognise, link with and, where feasible, reinforce other policy appraisal approaches used to shape development policies and programmes. This will help ensure environmental considerations are not overlooked and that SEA helps in underwriting the sustainability of their outcomes.

Three examples of other approaches are considered here: poverty and social impact analysis (PSIA); conflict/post-conflict and disaster assessment, and the diagnostic tool country environmental analysis (CEA). More information about PSIA and a range of other approaches to which SEA must relate, is in Annex B.

2.5.1. Poverty and social impact analysis (PSIA)

PSIA examines the distributional impact of policy reforms on the well-being or welfare of different stakeholder groups, and has an important role in the elaboration and implementation of poverty reduction strategies in developing countries. It has evolved out of awareness of the need to understand better the implications of policies recommended, and conditions required, by the IMF and World Bank in their lending programmes. The World Bank has made a specific commitment (Operational Directive 8.60) for when PSIAs will be undertaken. The practice so far has resulted in lender-led due diligence PSIAs aiming to verify externally to what extent PRSPs have had poverty reduction impacts.

Awareness of the benefits of PSIA, and of the need for them to be integrated into the decision-making process of recipient countries, is leading to a much wider understanding of PSIAs focused on evidence-based, pro-poor and inclusive policy-making. In some current cases (*e.g.* in the Balkans), the World Bank is addressing the linkages between environmental management and poverty as part of the PSIA.

PSIA has focused almost exclusively on economic, social, political and institutional analysis. Initially environmental considerations were put aside to allow the integration of the other methods and tools. Many of these tools are now used in SEA, or are likely to be relevant as SEA becomes more holistic. These tools are now well documented and there are increasing examples of good practice to draw on.[1] Whilst there has been good progress, there is a need to address environmental concerns more strongly to ensure the longer-term sustainability of proposed interventions. Progress in integrating SEA and PSIA will help towards a more sustainability-oriented form of impact assessment (more on PSIA in Annex B).

2.5.2. Conflict, post-conflict, and disaster assistance assessment

Development agencies are increasingly focusing on countries where poor governance is leading to a greater risk of conflicts that will undermine development efforts – sometimes known as fragile states. Their attentions are on both preventive strategies and supporting post-conflict situations to re-establish the foundations for sustainable development.

Tools are being developed to systematically assess conflict-related risks from different factors and to shape development in fragile states or those recovering from violent conflict.[2] There is a strong case for integrating environmental concerns when using these tools, particularly since environmental stress can contribute to conflict and is also a challenge to effective development interventions after conflict (see Box 2.3).

Strategic conflict assessment (SCA) was developed to help analyse conflict, assess conflict-related risks associated with development or humanitarian assistance, and develop options for more conflict-sensitive policies and programmes. It is a single exercise with three components: conflict analysis, analysis of donor responses and analysis of strategies and options.

The Drivers of Change (DoC) approach helps understand the dynamics of change and poverty reduction in developing countries. It focuses on the underlying and longer-term factors that affect the environment for reform and those that more directly affect the incentives and capacity for pro-poor change. DoC studies are starting to influence donor policy and highlight political and institutional issues in programme design across sectors.

It is important to recognise the linkages between environmental stress and conflict as part of the SCA approach, or indeed the DoC factors that link with access to natural

I.2. UNDERSTANDING STRATEGIC ENVIRONMENTAL ASSESSMENT

> **Box 2.3. Linkages between environmental stress and conflict**
>
> A recent review of the research on environmental stress and conflict revealed that environmental stress is a significant factor in either contributing to or aggravating conflict in many parts of the developing world.
>
> - Environmental stresses alone rarely lead to conflict. It usually contributes indirectly to conditions – political, social or economic – in society which result in or exacerbate conflict.
> - Where violence linked to environmental stresses happens, it will usually be sub-national and not inter-state.
> - Environmental stresses link with other factors that contribute to conflict, such as public health problems, or weaken a state's ability to prevent conflict.
> - The relationships are complex and the same conditions may lead to different outcomes depending on political and governance issues.
>
> *Source:* ERM (2002).

resources or reliance on natural resources as a source of public finance. There has been significant input on how environmental stress can contribute to the pre-conditions for conflict. This gives a framework for the integration of SEA with conflict assessment (see Box 2.4).

> **Box 2.4. SEA in post-conflict countries**
>
> SEA should only be applied when environment is a priority and when certain preconditions in the country are met. *Priorities* include circumstances:
> - Where environmental issues were, or may be, a source of conflict.
> - When badly planned reconstruction actions may seriously damage the environment.
> - When environmental programming could open peace-building opportunities that could not be better developed in other sectors.
>
> SEA will only be effective where an institution (usually the State) exists in a country that has the mandate, the capacity and the willingness to follow up on the key results of the actions agreed in the SEA, and when stakeholders are both willing and able to participate without risk. This means that, in most cases, SEA will not be a priority during the first stages of reconstruction that focus on relief, rehabilitation and institution building. It could be a priority, however, during the later stage of structural development and consolidation.
>
> *Source:* Verheem *et al.* (2005).

The purpose of an SEA in conflict-affected or post-disaster situations should be to help prevent natural resources from becoming a source of conflict. The SEA process should be designed to ensure that one side is not favoured over another, which would risk exacerbating divisions. It should aim to strengthen or restore natural resource-based livelihoods in resource-scarce settings, and to reduce opportunities for natural resource-based trade to fuel war economies. The SEA process should not jeopardise ongoing peace-building initiatives. It can be a relatively safe opportunity to bring disputants together over a shared concern with relatively low visibility (i.e. the environment), and thus contribute to peace-building.

The SEA process is not different in post conflict circumstances other than the need to 1) build greater sensitivity into the process regarding participation, notably to avoid putting participants at risk, and 2) focus particular attention on the drivers of the conflict, often competition over finite or diminishing natural resources.

Similarly, SEA can be applied to development assistance targeted at countries recovering from major natural disasters. In post-disaster situations, the goal of SEA is to help prevent further disasters where possible, or identify adaptation measures to mitigate the impact of potential future disasters. This can range from integrating environmental management priorities into the overall disaster relief strategy, to incorporating it in the more detailed environmental risk management applied to the immediate and continuing disaster relief efforts.

2.5.3. Country environmental analysis

Development agencies undertake policy level analysis of country environmental priorities, policy options and implementation capacity. SEA should take account of these reviews and fully integrate their findings. A key example is country environmental analysis (see Box 2.5). Typically, these studies take a broad view, rather than focusing on a particular

---

Box 2.5. Country environmental analysis

Country Environmental Analysis (CEA) is a flexible tool with three analytical building blocks: assessment of environmental trends and priorities; policy analysis; and assessment of institutional capacity for managing environmental resources and risks (World Bank, 2002a). It has three main objectives (World Bank, 2003):

- To facilitate mainstreaming by providing systematic guidance on integrating information on, and analysis of, key environment, development and poverty links into the country policy dialogue. The mainstreaming of environmental issues is more likely to happen when the diagnostic work is carried out, before preparing poverty reduction strategy papers (PRSPs), CASs, and large structural adjustment operations and other programmes.

- To guide environmental assistance and capacity-building supported by the World Bank or other development partners through an assessment of capacity issues, especially in relation to specific environmental priorities.

- To facilitate a strategic approach to environmental safeguard issues by providing information and analysis about environment-development links at the earliest stages of decision making. This will help shape key lending and programmatic decisions at the country and sectoral levels and help manage risks at the project level.

CEA provides a framework to systematically link country-level analytical work with strategic planning processes. Like other country-level diagnostic analyses, CEA is linked with a wide range of collaborative work with developing country and other development partners to guide their development assistance. Many of the tools and analytical approaches used in CEA approximate to those used in SEA, but their focus is large-scale and general. In an individual country, lessons can be drawn from previous SEA applications to provide key inputs to the broader review in the CEA. Conversely, the CEA can identify sectors and policies where a more in-depth analysis through SEA could provide more specific guidance for policy development.

country PPP. They define strategic priorities and bring social, economic and environmental issues together, or take a countrywide view of institutional capacity.

Such studies often provide inputs to broader development agency programming approaches including country assistance strategies or plans, which are the basis of a development agency's programme in a specific country (see Chapter 5). Hence, they often influence the content of an agency's support to a particular country, as well as lead to the further application of specific SEAs.

Notes

1. These documents are listed in the bibliography. Most of them are available on the World Bank PSIA Web site: *www.worldbank.org/psia*.

2. A Conflict Prevention and Post-Conflict Reconstruction Network (CPRN) has been established (*www.bellanet.org/pcia*) and a handbook on Post Conflict Impact Assessment (PCIA) prepared (Hoffman, undated *www.berghof-handbook.net*). OECD DAC's Fragile States Group has developed Principles for Good International Engagement in Fragile States. An SEA approach can influence these initiatives to ensure the environmental and natural resources linkages are not overlooked.

PART I

Chapter 3

The Benefits of Using Strategic Environmental Assessment in Development Co-operation

The application of SEA in the context of development co-operation has a range of benefits – both improvements in process and improved development outcomes. Inevitably, process benefits are easier to identify and document, but this should not obscure the improved outcomes that result in ensuring environment is mainstreamed in strategic decisions. This chapter provides examples of how SEA has improved the process of decision taking, making important contributions to development effectiveness.

> ### Box 3.1. **SEA benefits at a glance**
>
> - SEA can **safeguard the environmental assets and opportunities** upon which all people depend, particularly the poor, and so promote sustainable poverty reduction and development.
> - SEA can **improve decision making** related to policies, plans and programmes, and thus **improve development outcomes** by:
>   1. Supporting the integration of environment and development.
>   2. Providing environmental-based evidence to support informed decisions.
>   3. Improving the identification of new opportunities.
>   4. Preventing costly mistakes.
>   5. Building public engagement in decision making for improved governance.
>   6. Facilitating transboundary co-operation.

3.1. Supporting the integration of environment and development

3.1.1. *SEA and poverty reduction strategies and related national planning strategies*

The development co-operation community is moving towards strategic policy frameworks, such as poverty reduction strategies, as the platform for development assistance (including direct budget support and sector support programmes). SEA is a means to ensure the integration of environmental (as well as social and economic) linkages in the design of poverty reduction strategies so that they result in better and more sustainable development through improved contribution of environment and natural resources to poverty reduction.

SEA is not an obstacle to programme approval. Rather it provides a process for integration and programme improvement that will give greater confidence to the decision maker, particularly where development resources are under stress and poverty reduction needs are great. Tanzania's second poverty reduction strategy included several SEA elements (see Box 3.2).

> Box 3.2. **Incorporating environmental considerations in Tanzania's second PRS**
>
> **Issue**
>
> When developing its second poverty reduction strategy, *the National Strategy for Growth and Reduction of Poverty* (NSGRP), Tanzania used an approach incorporating key SEA elements. These included an extensive consultative review process, a systematic and integrated assessment to incorporate environment in sector policies, in the national budget and in the poverty reduction support credit.
>
> **Key benefits**
>
> - The NSGRP has a specific goal on environment sustainability and 14% of the targets are directly related to the environment and natural resources.
> - Interventions on environment are expected to make significant contributions to other targets focusing on growth, health and governance.
> - A set of poverty-environment indicators has been developed as part of the national poverty monitoring system that will be used to assist on reporting on the MDGs.

3.1.2. *SEA and other policy-level reforms*

Development agencies frequently provide finance for proposed changes in legal systems, sector development policies, capacity-building and policy reforms at various levels. While the provision of consultant services to provide advice does not, in itself, raise particular issues, the recommendations arising from such technical co-operation projects in relation to, for example, legal reform or agriculture policy may raise significant environmental issues. All development agency proposals for capacity-building, policy and legal or sector reform should be evaluated to see if there are associated environmental or social implications that may result from changes. If so, an SEA could be an effective way to identify and evaluate alternatives and propose recommendations to take into account in the implementation of the policy advice. These recommendations would improve the effectiveness of the technical co-operation efforts and help prevent unexpected negative outcomes of the work.

3.2. Identifying unexpected potential impacts of reform proposals

Advice on restructuring a railroad network, for example, may not appear to be associated with environmental issues. However, efficiency considerations may call for certain linkages to be closed or by-passed. This may require local farmers to transport produce to markets over longer distances and increase the impacts on regional roads and associated pollution. An SEA could evaluate the environmental and social costs and benefits of alternatives, outline the trade-offs that need to be considered, and make recommendations.

3.3. Improving the identification of new opportunities

By encouraging a systematic examination of development options, SEA can help decision makers to identify new opportunities, thus avoiding the missed opportunities often associated with limited choices. SEA can also help policy-makers in different but closely related sectors to identify win-win opportunities. In Ghana, for example, an SEA of the *National Strategy for Growth and Poverty Reduction* helped identify new opportunities in the forestry sector while safeguarding water resources (see Box 3.3).

Box 3.3. Achieving positive revision to forest policies in Ghana

Issue

An examination of the Ghana Poverty Reduction Strategy (GPRS) identified potential conflicts between the forest policy (aimed at broadening the resource base of the wood industry) and environmental protection of river system bank-side ecosystems. As a result, Ghana's forest policy was modified. In less than six months, the government had set up nurseries to raise bamboo and rattan plants to increase the supply of raw materials for the industry, thereby helping protect riverbanks from uncontrolled harvesting of wild bamboo and rattan.

Key benefits

- Reduced pressure on primary forests and fragile river ecosystems.
- Creation of new timber resources.
- Employment.

Box 3.4. Thermal Power Generation Policy, Pakistan: an early SEA would have helped

Issue

In the mid 1990s, in response to rapidly expanding industrial activity and an increasing population, Pakistan's Government decided to stimulate increased power generation. The Independent Power Plants Policy provided incentives for investments in thermal power generation. No SEA was made; instead investors had to submit an EIA without considering potential cumulative effects. Investors were given the freedom to choose the site, the technology and the fuel and many of these plants were installed with little or no pollution control devices. Leading energy experts and the Water and Power Development Authority raised objections to the policy but were ignored. EIA was used as a down-stream decision-making approach applicable to individual projects, especially after deciding the site, technology and fuel. As a consequence, many thermal power stations using high-sulphur furnace oil became clustered in one city and added to the already polluted air. Alternatively, they were developed in a scattered way in remote places, which made it difficult to connect them with the national grid system.

Key costs

- Increased pollution
- Relocation of plants – following public pressure and lobbying – at considerable cost.
- Delayed delivery of energy.

Later, IUCN Pakistan reviewed the policy through an SEA-like process. This made it clear that EIA alone was not sufficient. Following a training programme, the Planning and Development Department began to request SEAs for major national and provincial-level initiatives at the policy level. The Independent Power Plants Policy is still widely cited as an example of a (bad) policy which, had it been subject to an SEA, could have been changed and major environmental and economic losses to the country avoided.

Source: Naim (1997a, 1997b and information from IUCN Pakistan).

3.4. Preventing costly mistakes

SEA can prevent costly mistakes if it alerts decision makers about unsustainable development options at an early stage in the decision-making process (see Box 3.4). The costs involved might consist of unbudgeted time and resources in handling disputes with local communities or mitigation of avoidable harm through pollution. In extreme cases, it may be necessary to relocate or redesign facilities. Clearly, the costs of necessary reparation, mitigation or duplication of investment are counter-productive to the aims of development assistance.

3.5. Building public engagement in decision making for improved governance

Public engagement is critical. A policy reform or programme will be far more effective when the values, views, opinions and knowledge of the public are reflected in the decision-making process. For the decision maker, effective public engagement will impart a higher degree of confidence in reaching a decision, and will lower the risk of a decision that could lead to unfavourable results (see Box 3.5).

Box 3.5. Sector EA of Mexican Tourism

Issue

Tourism accounts for approximately 9% of Mexico's GDP. It is the country's third largest source of foreign currency (US$10 800 million a year), drawing more than 52 million domestic and 20 million international visitors in 2004. However, if de-linked from sustainable planning and investment, tourism growth can threaten the very resource on which it is based. In a 2002 tourist survey, environmental quality – one of the key determinants for selection of tourist destinations – received the lowest rating.

The 2001-06 National Development Plan emphasised the need for economic development with human and environmental quality. An SEA process of the tourism sector was initiated to formulate and implement a sustainable policy for the country. To ensure broad participation and commitment sectors, a high-level mechanism for inter-institutional co-ordination (the Inter-sectoral Technical Working Group, ITWG) was established. The group comprised representatives from tourism, environment, forests, water, urban development, and the interior and finance ministries. It set sector priorities, an action plan for implementation, and medium-term monitoring indicators. More recently, this group has been formalised as the Inter-sectoral Commission for Tourism.

Key benefits

- Provided environmental-based evidence to support informed decisions. The SEA identified environmental opportunities and constraints associated with different growth scenarios, as well as sector and environmental priorities consistent with optimising the benefits from tourism without over-exploiting the environment.

- Participation from all sectors and relevant stakeholders. The ITWG enabled parties with different mandates over natural resources and other issues to make durable commitments and reach agreements with a long-term perspective.

- The findings of the analytical work are informing a policy for the sustainable development of tourism.

Source: World Bank (2005).

> **Box 3.6. SEA for Water Use, South Africa**
>
> In 2000-04, the South African Department of Water Affairs and Forestry (DWAF), with support from UK DFID, undertook a pilot "SEA for Water Use Study" in the Mhlathuze Catchment in KwaZulu Natal. This found that:
>
> - The catchment was under water stress and there was no surplus for allocation to new users.
> - There was a deep historical inequity in the allocation of water resources between established commercial sectors and the community, although more that half of the land was in communal ownership and occupied by 80% of the population in the catchment.
> - There were prospects for a more equitable sharing of this water.
> - Putting the "haves" and the "have-nots" together to debate needs, demands and visions face-to-face was essential to mutual understanding and suggested that ways could be found to ameliorate inequities.
>
> The SEA was extended to the entire Usutu-Mhlathuze Water Management Area (seven large catchments) but struggled to isolate and resolve issues at this scale because of the disparate and disconnected nature of these catchments. The scale of the Mhlathuze catchment (4 000 km^2), involving a single water source, proved conducive to both communication and problem solving.
>
> DWAF has now embraced SEA as an approach for use in catchment planning and management both for addressing sectoral issues, for large-scale projects, and at catchment scale. SEA is a recognised tool supporting the implementation of the National Water Act (1998), and the principles of SEA widely adopted into planning and decision-making approaches, with the following specific objectives:
>
> - To ensure best use of water in an integrated way to most benefit society and the economy without degrading the environment.*
> - To encourage people to become involved in catchment affairs and to link users with decision makers; to assess and analyse data from the catchment.
> - To provide decision makers with reliable data from the catchments for more informed decisions.
>
> * The National Water Act promotes the beneficial, efficient and sustainable use of water in the public interest.
> Source: www.dwaf.gov.za/sfra.

SEA can benefit when *a)* those with particular and relevant knowledge regarding the proposed policy or programme, and *b)* those that could be positively or negatively affected by the decision, are invited into the SEA process.

SEA supports *good governance* by:

- Encouraging stakeholder participation in decision making.
- Increasing transparency and accountability in decision making.
- Clarifying institutional responsibilities (*e.g.* division of responsibility between local government, line departments, state/provincial and national/central governments).

3.6. Facilitating transboundary co-operation

SEA can provide an important arena for regional co-operation, *e.g.* to address difficult issues concerning shared resources such as waterways; upstream actions which have downstream effects; pollution impacts across boundaries; transfrontier protected areas, transport connections, infrastructure, migration (see Chapter 5, *Guidance Note 10* for a case example on transboundary assessment of the Nile Basin).

3.7. Safeguarding environmental assets for sustainable development and poverty reduction

SEA enhances the prospects of safeguarding the environment and the natural systems that are the critical foundations for human health and livelihood. The world's poor depend most directly and heavily on natural resources both for subsistence and income opportunities. MDG 7, aiming to ensure environmental sustainability, is a cornerstone on which strategies to reduce poverty must be built. Yet the reality is that the "environmental assets of poor households are under severe and increasing stress".* When applied as part of development policy development and plan making, SEA offers a systematic process to avoid or minimise adverse impacts on the environment and to enhance resource opportunities (see Box 3.6).

* UNDP/UNEP/IIED/IUCN/WRI 2005.

ISBN 92-64-02657-6
Applying Strategic Environmental Assessment
Good Practice Guidance for Development Co-operation
© OECD 2006

PART I

Chapter 4

Towards Strategic Environmental Assessment Good Practice: Principles and Processes

This chapter sets out guiding principles and key generic steps for SEA as they have emerged from practice.[1] Readers familiar with SEA may wish to go straight to Chapter 5 which outlines more specific guidance on applying SEA in development co-operation. SEA always demands a great focus on strengthening institutional capacity to take environmental issues into consideration, particularly when applied to policies (see Section 4.2). Where SEA is applied to plans and programmes, a more structured approach to integrating environmental considerations can be used, adapting the characteristic steps of EIA (see Section 4.3). In policy making, usually this will not be possible because of the complex, non-linear character of this process.

4.1. Basic principles for SEA

To be influential and help improve policy-making, planning and decision-taking, an SEA should:

- Establish clear goals.
- Be integrated with existing policy and planning structures.
- Be flexible, iterative and customised to context.
- Analyse the potential effects and risks of the proposed PPP, and its alternatives, against a framework of sustainability objectives, principles and criteria.
- Provide explicit justification for the selection of preferred options and for the acceptance of significant trade-offs.
- Identify environmental and other opportunities and constraints.
- Address the linkages and trade-offs between environmental, social and economic considerations.
- Involve key stakeholders and encourage public involvement.
- Include an effective, preferably independent, quality assurance system.
- Be transparent throughout the process, and communicate the results.
- Be cost-effective.
- Encourage formal reviews of the SEA process after completion, and monitor PPP outputs.
- Build capacity for both undertaking and using SEA.

In designing effective SEA approaches, practitioners need to be aware of the following:

- Strategic planning is not linear, but a convoluted process influenced by interest groups with conflicting interests and different agendas; it is therefore important to look for "windows of opportunity" to initiate SEA during cycles of the decision-making process.
- Relationships between alternative options and environmental effects are often indirect; so they need to be framed in terms relevant to all stakeholders (*e.g.* politicians, government agencies and interest groups). One way of doing this is by linking environmental effects to their specific policy priorities.

- Strategic issues cannot be tackled by a one-off analysis; they need an adaptive and sustained approach as strategies and policy-making take shape and are implemented.
- The value of SEA in strategic planning depends greatly on capacity within the responsible authorities to maintain the process and act on the results.

4.2. The institutional dimension of SEA

Effective SEA depends on an adaptive and continuous process focused on strengthening institutions, governance and decision-making processes rather than just a simple, linear, technical approach focussed on impacts, as is often found in EIA. This is of particular importance in the context of policy-level SEA (see Box 4.1). Figure 4.1 indicates the key steps to strengthen institutional capacity. These are discussed in more detail in the following paragraphs.

Figure 4.1. **Steps to address institutional considerations in SEA**

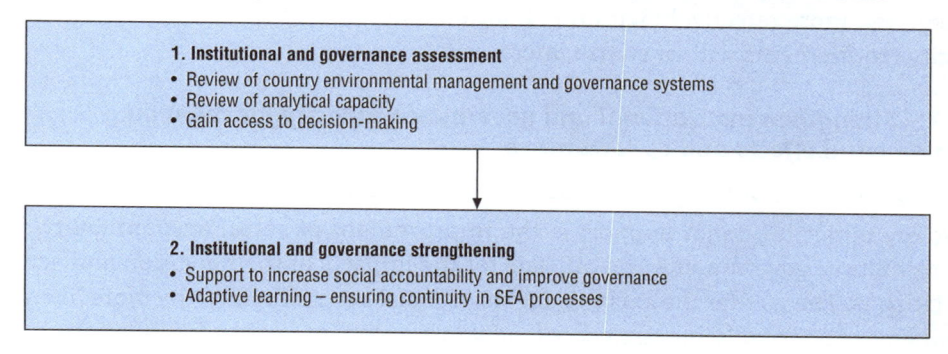

1. Institutional and governance assessment
- Review of country environmental management and governance systems
- Review of analytical capacity
- Gain access to decision-making

2. Institutional and governance strengthening
- Support to increase social accountability and improve governance
- Adaptive learning – ensuring continuity in SEA processes

Step 1 – Assess institutional capacities to manage effects and opportunities

◆ *Review country environmental management and governance systems*

Given the challenges of applying SEA to PPP formulation and reform, it is essential to assess the systems which are in place to address the environmental linkages with key policy goals and issues. This includes, in particular, assessing how well equipped a country's institutional capacity is to manage uncertain or unexpected environmental impacts and to take advantage of environmental opportunities. There is significant experience among donors and partner governments in undertaking institutional assessments. Country Environmental Analysis (CEA) (see Box 2.5) is a current example and can be an effective entry point provided the emphasis is not limited to environmental institutions and capacities, but makes the link with economic and social institutions.

The review of country systems should not be limited to government environmental agencies, but should also tackle the institutions, incentives and processes that support improved governance and public and private sector engagement, particularly to promote responsible environmental and social management. It should also examine a country's environmental governance mechanisms for ensuring and reinforcing social accountability, *e.g.* people's access to the judiciary to address environmental pollution or natural resource allocation issues, or dissemination of information in a manner that is easily interpretable to allow communities to play a role as informal regulators.

✦ Review analytical capacity available to PPP-making institutions

The success of SEA hinges on country-based analytical capacity. There will generally be a range of analytical capacities in government, research and academic institutions, civil society organisations (CSOs) and the private sector. Most will have focused on impact assessment approaches and some will have engaged in wider analytical frameworks relevant to policy processes (*e.g.* state of the environment reports and sector studies). Links should also be made with efforts to integrate other forms of impact analysis (*e.g.* PSIA) into the institutional structure.

✦ Take opportunity to gain access to the decision-making process

SEA will often require an opportunistic approach. The champions' of SEA, whether from the donor community or government, need to respond when windows of opportunity arise for mainstreaming environmental issues in policy formulation. Also, where possible, past success should be built on and experience used to assess institutional capacity. As SEA is integrated more effectively within PPP-formulation and decision making, institutional capacity requirements will, of course, increase.

Step 2 – Strengthen institutional and governance capacity for managing environmental effects and opportunities

✦ Support mechanisms that increase social accountability and improved governance

A key target for donor support is the improvement of social accountability, *i.e.* the responsibility of governments and officials for the impacts of their decision and actions to their citizens. The greater the existing degree of social accountability, the more likely it will be that environmental issues are successfully integrated into policy formulation.

Social accountability can be increased by focusing on electoral processes, legal and judicial reforms, independent audits and oversight processes, and access to information. All efforts to increase the rights of the citizens and hold governments and officials accountable are likely to lead to improved governance and greater transparency. An additional element is support to CSOs to enable them to be more effective in the policy dialogue and to increase their analytical capacity.

With better governance comes greater integration of environmental issues with social and economic policy goals. The public have greater opportunities to challenge policy makers to address environmental issues and to be more transparent about the environmental implications of economic and social policies.

✦ Assist countries to ensure continuity of SEA processes

The importance of sustained assistance to countries engaging in SEA can hardly be overstated. SEA should not be a once-off event that results in a discrete output – but an institutional process that adapts to the momentum and cyclical nature of policy formulation. Therefore, the need for capacity-building is of great significance. It should focus on analytical, participatory and political requirements, and on adaptive learning to capture lessons from effective processes and institutional arrangements (see Chapter iii).

Assistance needs to be sustained to be effective. Planning processes and capacity building have medium- to long-term time frames. The focus should be on building constituencies as well as public administration capacities.

> ## Box 4.1. **Institutions centred SEA**
>
> The complex interactions between political, social and environmental factors create special challenges for the environmental assessment of policies. For example, the formulation of macroeconomic and sectoral policies – while driven by the objective to promote public well-being – is also subject to intense political pressures from different stakeholders with sometimes conflicting interests. In the context of weak institutional and governance frameworks, powerful stakeholders and elites often prevail over other stakeholders, including groups such as local and indigenous communities who may be particularly vulnerable to the social and environmental impacts of the policy choices made.
>
> As compared to project-level EIA, the SEA of policies therefore requires a much more thorough understanding of political economy factors and institutional settings. This includes, in particular, recognising that differences in political power among affected stakeholders imply significant differences with regard to bargaining power and ability to influence policies and, ultimately the economic, social and environmental impacts of policy decisions.
>
> To address these challenges, the World Bank is currently testing and validating an "institutions-centred approach to SEA". This approach acknowledges that strategic decision making at the higher levels is heavily influenced by political factors and accordingly focuses on institutional and governance dimensions.
>
> Unlike the traditional approach on "impact-centred SEA", the institutions-centred approach to SEA does not focus primarily on assessing potential impacts and mitigating these impacts. Rather, it places special emphasis on improved governance and social accountability (i.e. the obligation of public officials and decision makers to render account towards their citizens, and the society at large, regarding their plans of actions, their behaviour and the results of their actions) on a continuous basis as well as social learning, which is essential in order to raise more attention to environmental issues and to continuously improve the design of public policies.
>
> **Key steps**
>
> A first step is the identification of environmental priorities and assessing the advantages and disadvantages (through, for example, tangible and intangible benefits and costs) of different courses of actions with respect to the proposed policy on these priorities. This requires rigorous consideration of key environmental problems and risks within a country, sector or region, including an assessment of the underlying causes of environmental stresses. In addition, an analysis of the winners and losers for each possible course of action is important. Intersectoral coordination is also critical as policies are likely to have multi-sector environmental effects. Effective and sustained public engagement is vital for policy-level SEA. Stakeholders' perceptions, in particular of the possible losers, of the potential effects of a policy therefore need to be incorporated and validated using the best available evidence possible, taking into account their relative environmental vulnerability and power to influence the policy process, through the use of stakeholder analysis tools. In addition, a number of tools or techniques such as comparative risk assessment, cost of environmental damage studies, survey-based and participatory assessments can be used to prioritize an environmental issue and how it can be affected by the proposed policy.
>
> The second step involves assessing the country environmental management systems to address the effects of policies on the identified environmental priorities. If country environmental management capacity is inadequate, the third is establishing the institutional and governance strengthening requirements to address these effects effectively. (These steps are discussed in more detail in Section 4.2.)
>
> *Source:* Integrating Environmental Considerations in Policy Formulation. Lessons from Policy-Based SEA, Report No. 32783, World Bank, 2005.

4.3. Stages and steps for undertaking SEA at the plan and programme level

Many countries and agencies have developed SEA guidelines and procedures. So far, these are mainly aimed at strengthening plan and programme development and are based on an adaptation of the steps characteristic for EIA. Practical experience with these approaches suggests that good practice SEA should involve **four stages** (see Figure 4.2).

Each stage can be further subdivided into steps/tasks (indicated by the arrows in the text) But these do not need to be carried out in sequence.[2]

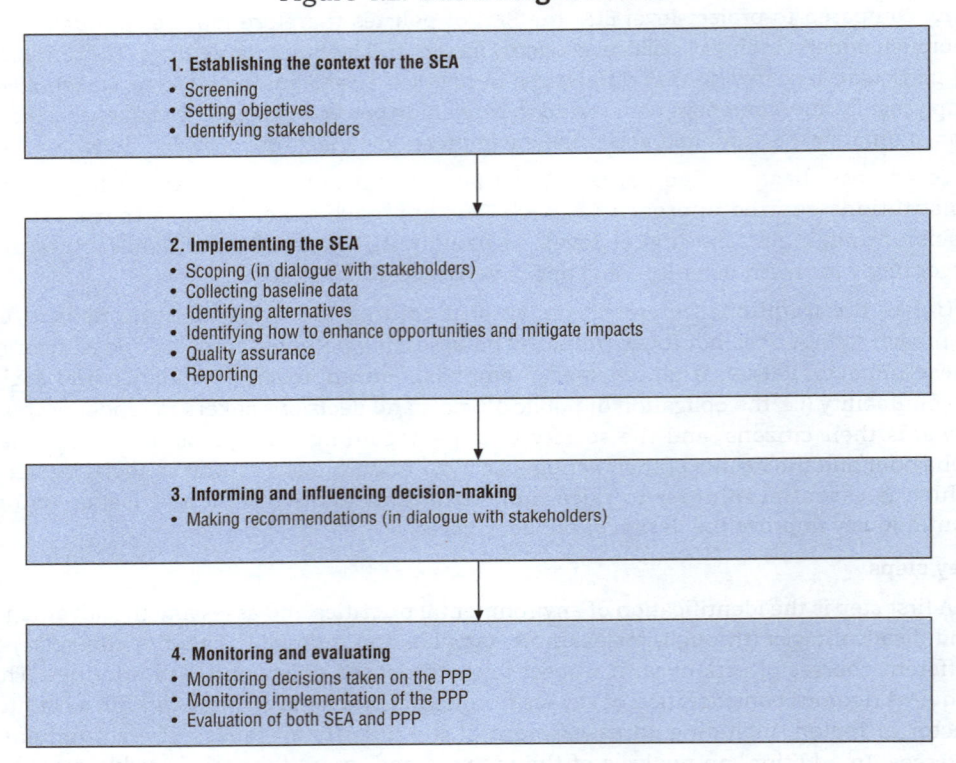

Figure 4.2. **Basic stages in SEA**

1. **Establishing the context for the SEA**
   - Screening
   - Setting objectives
   - Identifying stakeholders

2. **Implementing the SEA**
   - Scoping (in dialogue with stakeholders)
   - Collecting baseline data
   - Identifying alternatives
   - Identifying how to enhance opportunities and mitigate impacts
   - Quality assurance
   - Reporting

3. **Informing and influencing decision-making**
   - Making recommendations (in dialogue with stakeholders)

4. **Monitoring and evaluating**
   - Monitoring decisions taken on the PPP
   - Monitoring implementation of the PPP
   - Evaluation of both SEA and PPP

Stage 1: Establishing the context for the SEA

✦ *Review the need for the SEA, and initiate preparatory tasks*

An early step in the SEA process is "**screening**" to decide whether an SEA is appropriate and relevant in relation to the development of a PPP in the area under consideration. Integral to this will be **establishing the objectives** of the SEA: how does it intend to improve the planning process; what is its role?

SEA is designed to explore and evaluate suitable alternatives. The sooner an SEA is introduced to policy formulation and plan-making, the greater the chances are to identify opportunities and influence outcomes.

When it is decided that an SEA is appropriate, it is important to secure governmental support. The explicit focus throughout the subsequent process should be on integrating environmental considerations (alongside economic and social ones) into key decision-

> ### Box 4.2. **Preparatory tasks in SEA**
>
> - Establish the terms of reference. These should apply the basic principles of SEA (see Section 4.1).
> - Set up a management team/steering committee and appoint an SEA co-ordinator/manager.
> - Clarify and confirm the specific goals and objectives of the SEA in relation to the objectives of the PPP with partners and stakeholders.
> - Develop capacity-building and a communication plan for the SEA.
> - Determine if the objectives of the PPP are in line with existing (environmental or other) objectives of country/region/sector authorities.
> - Set appropriate decision criteria from these objectives and the broader development agendas of the parties.
> - Set definite and realistic timescales.
> - Agree the required documentation.
> - Confirm sources of funding.
> - Announce the start of the planning process; bring key stakeholders together to agree on problem, objectives, alternatives and measures for quality control.
>
> **Special tasks in development co-operation**
>
> - Ensure full account is taken of the development priorities of the developing country.
> - Ensure the appointments to the SEA team are made – whether in-house – preferably engaging national expertise, through local consultants supported by technical assistance from international consultants, or as a partnership venture as necessary.
> - Determine whether other institutions (including donors) have carried out, or intend to carry out, an SEA relevant to the PPP in question and, in such circumstances, seek to engage in a joint process. This might involve:
>   - Delegating the SEA to another partner (including donor) with better experience in the area/country; and agreeing shared financing.
>   - Alternatively, carrying out the SEA with the host country on behalf of other partners (including donors).
>   - Joining staff and financial resources to undertake a common multi-partner (including donor) SEA process.
> - In parallel to seeking such a harmonised approach to SEA, it is crucial to integrate the SEA process with existing planning and assessment systems in the partner country and develop links with other impact assessment approaches in use.

making points when options and proposed activities are being developed and evaluated. A number of preparatory tasks are necessary (see Box 4.2).

✦ Identify interested and affected stakeholders and plan their involvement

SEA is a participatory process. It allows civil society, including the private sector and relevant stakeholders that will be affected by the proposed PPP, to contribute inputs to strategic decision making. Therefore, screening should include careful **stakeholder analysis** to identify stakeholders and prepare a communication plan to be used throughout the SEA. If the public is not used to being engaged, particularly at the strategic level, and if there are no precedents, it is critical to include an education component in the public engagement process. Active public engagement should take place from Stage 2 onwards to the review of the draft SEA report.

A public engagement and disclosure plan can assist in identifying relevant stakeholder groups and appropriate communication methods. It is important to identify and engage those stakeholders who are the most exposed to environmental degradation. In general, environmental pressures tend to affect the poor and vulnerable sections of the population more seriously. To ensure that all relevant knowledge is drawn on, both women and men should be included in this process.

Stage 2: Implementing the SEA

+ *Determine the scope of the SEA*

A **scoping** process should establish the content of the SEA, the relevant criteria for assessment (*e.g.* goals set out in the National Sustainability Development Strategy). These should be set out in a scoping report. A pragmatic view needs to be taken on how much can be achieved given the time-scale, available resources, and existing knowledge about key issues. An open and systematic process should be followed. The SEA should actively engage key stakeholders to identify significant issues associated with the proposal and the main alternatives. Based on these issues, and the objectives of the SEA, decision criteria and suitable indicators' of desired outcomes should be identified. Scoping may also recommend alternatives to be considered, suitable methods for analyses of key issues and sources of relevant data.

Scoping procedures and methods, such as matrices, overlays, and case comparisons, can be used to establish cause-effect links between different specific plans or programmes or to identify the environmental implications of more general policies or strategies. A detailed options review may be undertaken as part of the scoping process to clarify the environmental advantages and disadvantages of different potential courses of action. Scoping meetings with stakeholders should result in a revision of the scope or focus of the SEA and improvements (as needed) to the draft engagement plan developed during screening.

+ *Establish participatory approaches to bring in relevant stakeholders (as part of the Scoping process)*

As noted above, effective and sustained public engagement is vital for effective SEA. By their very nature, PPP decisions are embedded in the political domain and involve political dynamics – including the engagement of the stakeholders who are likely to be most affected or who are most vulnerable. Understanding the power relations between different stakeholders, and how they interact with each other and the environment, are essential for good analysis and process management.[3]

One of the challenges is to ensure that public engagement is meaningful and not just a case of providing detailed, rigorous and comprehensive information. The engagement process must provide an opportunity to influence decisions. Stakeholders groups identified as most affected by a given PPP may be politically and or socially marginalised and have little or no prior experience in providing input to decision making.

Public consultations processes will have to identify the best means to ensure that they can participate effectively and their viewpoints are given proper consideration. This may involve, in particular, reaching stakeholders who may not have access to the internet, lack access to public libraries, speak a different language, are illiterate, have cultural differences or other characteristics that need to be taken into consideration when planning for their engagement.[4]

Depending on the nature of the country's political institutions and processes, there will be a need to integrate any SEA process with the public engagement process as a whole, or to adopt other approaches where needed. Also, public engagement needs to be sustained, structured, and co-ordinated with the phases of formulating and implementing PPPs – emphasising equally the positive contributions and harmful effects.

✦ Collect baseline information

SEA needs to be based on a thorough understanding of the potentially affected environment and social systems. This must involve more than a mere inventory, *e.g.* listing flora, fauna, landscape and urban environments. Particular attention should be paid to important ecological systems and services, their resilience and vulnerability, and significance for human well-being. Existing environmental protection measures and/or objectives set out in international, national or regional legislative instruments should also be reviewed.

The baseline data should reflect the objectives and indicators identified in the "scoping report". For spatial plans, the baseline can usefully include the stock of natural assets including sensitive areas, critical habitats, and valued ecosystem components. For sector plans, the baseline will depend on the main type of environmental impacts anticipated, and appropriate indicators can be selected (*e.g.* emissions-based air quality indicators for energy and transport strategies). In all cases, the counterfactual (or no-change scenario) should be specified in terms of the chosen indicators.

✦ Analyse the potential effects of the proposals and any alternatives

Identifying the potential direct and indirect or unintended effects of policy proposals and decision-making processes, as well as options for, and alternatives to PPPs is naturally more difficult than in the case of specific projects. The range of options or variables under consideration is often harder to define with certainty because the transmission channels through which effects may be experienced may be very complex, involving many aspects which are difficult to predict and analyse. This makes the indirect effects of paramount importance in the assessment. Certain measures can help to frame this issue, for example, the use of best *versus* worse case scenarios. Cumulative effects present particular challenges and may require expert consideration.

There is no single best method for impact analysis. Approaches should be selected that are appropriate to the issues at stake. The identification and evaluation of suitable options may be assisted by future "scenario building" and "back-casting methodologies".

Establishing the linkages with key economic and social policy goals requires a wide analytical framework, elements of which may already exist. For example, there may already have been a rigorous examination of the key environmental problems and risks within a country or region, including an assessment of the underlying causes of environmental stresses. If not, a partial analysis should be undertaken relevant to the scale or scope of the policy in question in order to assess the potential linkages between the environmental effects of the policy being assessed and key policy goals (*e.g.* in many poor countries, policies indirectly leading to rural environmental stress can impact negatively on poverty levels).

Assessment of the priority of such linkages and issues will reflect the perceived value of the environmental issues to the country. Such assessment can draw on a number of tools or processes *e.g.* comparative risk assessment, economic assessment of environmental damage, and survey-based and participatory assessments). They can be used to find objective measures of how important an environmental issue is, and thus how it should be factored into the policy formulation process alongside other issues.

✦ Identify measures to enhance opportunities and mitigate adverse impacts

It is important to focus on realising the positive opportunities of the planned activities and minimising any negative risks. Opportunities will generally enhance achievement of the MDGs and other development challenges. The aim is to develop "win-win" situations where multiple, mutually reinforcing gains can strengthen the economic base, provide equitable conditions for all, and protect and enhance the environment. Where this is impossible, the trade-offs must be clearly documented to guide decision makers.

A mitigation hierarchy should be followed for identified negative impacts: first avoid; second reduce; and third offset adverse impacts – using appropriate measures. Caution should be exercised if the analysis indicates a potential for major, irreversible, negative impacts on the environment. Often this may suggest selecting less risky alternatives. For less-threatening situations, standard mitigation measures can be used to minimize an adverse impact to "as low as reasonably practicable" (ALARP level).

Once mitigation has been taken into account, the significance of residual adverse impacts can be evaluated. This is an important measure of the environmental acceptability of the proposal; it is usually carried out against selected environmental objectives and criteria.

Examples of policy reforms with clear environmental implications include privatisation, energy policy, land reform, trade incentives, water supply and pricing. Table 4.1 shows how policy reforms in a variety of sectors can have positive and negative environmental consequences, and gives examples of measures that can be taken to enhance or mitigate them.

✦ Draft report on the findings of the SEA

Once the technical analysis is completed, the results and rationale for conclusions need to be reported. While a technical report may be necessary, it must be presented in an understandable format and appropriate language(s). This will often require short summaries and graphic presentations rather than a long report. A succint, non-technical summary should be included. This will be of particular use in explaining the findings to civil society, which needs to be well informed in order to submit comments.

✦ Provide an independent evaluation/review (quality control check) on the SEA

Designing an SEA to include the steps and practices outlined in Stages 1-3 will provide a basic level of process quality. However, specific measures of quality control assurance might be warranted, *e.g.* to ensure the credibility of the assessment in the eyes of all stakeholders. These measures will depend on the nature, context, needs and timeframe of the specific strategic initiative. Options to consider include:

- An independent review of SEA by experts or academics.

Table 4.1. **Examples of policy reforms and potential environment linkages**

Policy area	Reform	Potential environmental benefits	Potential environmental risks	Measures to enhance environmental benefits and mitigate risks
Energy	Fuel price reform, removal of subsidies.	Reduced emissions through increased production and consumption efficiency.	Removal of subsidies could lead to increased demand for fuel wood.	Property right reforms might be used to mitigate against deforestation in search for fuel wood.
Agriculture	Land reform.	Strengthened property rights generally improve management of natural resources.	Shrinking common property resources are overused by landless.	Ensure that the interest of the landless are considered. Provide training on fertiliser and pesticide use.
Private sector development	Business climate issues, taxation and protection of property rights, privatisation.	Increased competition and use of price signals generally improve resource use efficiency.	Weak legal environmental framework and unclear liabilities can lead to over exploitation of natural resources and high pollution levels.	Ensure adequate legal framework, monitoring and enforcement.
Tax reform	Tax incidence (income, assets, corporation, consumption); tax rates; exemptions; deductions.	Changes in prices due to tax reform can have powerful effects on household and corporate behavior. Natural resources positively/negatively affected depending on the reform. Subsidy removal generally has positive effects on natural resource use.	See benefits.	Environmental fiscal reforms where taxes on polluting inputs such as energy and resource royalties are used can lead to internalised environmental costs, increased resource efficiency and tax incomes.
Decentralisation	Decentralisation of power to regional or local administration. Reforms aim at increasing the efficiency of service delivery, accountability.	Accountable and representative local institutions can improve the management of natural resources.	Poor capacity to deal with environment and natural resource related issues. Risk that local elites exploit local natural resources (if no state vigilance). Capacity-building to strengthen local and regional administration.	Capacity-building to strengthen local and regional administration.
Trade	Trade reform	Increased competition may lead to improved resource use efficiency. Benchmarking of environmental performance standards by in-migrating industry.	Expansion of monocultures. Increased use of fertilisers and pesticide. Increased pressure to convert forests or wetlands to agriculture. Increased water and air pollution from industry	Improve environmental legislation to avoid becoming a "pollution haven".

For further reading see OECD (2005), and WRI/UNDP/UNEP/World Bank (2005).

- Internal audits by the Ministry of the Environment; "sounding boards" or steering committees consisting of representatives of key stakeholders.
- An independent expert commission.

✦ Public engagement in reviewing the draft SEA report

While public engagement should have been included at all appropriate stages (see paragraphs 94-97), the draft SEA report is a key stage and should be publicly available for a period of time agreed during the scoping stage. If meetings are held for public comment, smaller, focused meetings may be preferable to ensure adequate time for comment, rather than larger meetings where few people have the opportunity to speak. There is a variety of ways to gather opinion from the more vulnerable groups and ensure that they can meaningfully participate, *e.g.* surveys, interviews and meetings. Financial support, transport and food may need to be provided so the most marginalized can participate. An understanding of the political economy of the decision-making process, and the various responses from the stakeholder analysis, should suggest how to ensure effective consultation and influence on decisions.

✦ Prepare final SEA report

Typically, this would include sections/chapters on:

- The key impacts for each alternative.
- Stakeholder concerns including areas of agreement and disagreement, and recommendations for keeping stakeholders informed about implementation of recommendations.
- The enhancement and mitigation measures proposed.
- The rationale for suggesting any preferred option and accepting any significant trade-offs.
- The proposed plan for implementation (including monitoring).
- The benefits that are anticipated and any outstanding issues that need to be resolved.
- Guidance to focus and streamline any required subsequent SEA or EIA process for subsidiary, more specific undertakings such as local plans, more specific programmes and particular projects.

Stage 3: Informing and influencing decision making

✦ Making recommendations to decision makers

Presentation of the draft and final reports are important to influence key decisions. A clear, understandable and concise Briefing Note or Issues Paper can help to ensure that decision makers are fully aware of key environmental issues linked to the PPP. From the outset, through steering committees, other structures and public engagement mechanisms, decision makers and stakeholders have opportunities to shape the outcome of the SEA, *e.g.* identification of issues, choice of indicators, scope of work, and selection and evaluation of proposed development options and alternatives.

It is often a learning process for authorities and civil society to work together on a PPP. Decision makers need to know the options open to them, what the likely effects of choices are, and what the consequences would be if they failed to reach a decision. This information should be clearly set out in the advice given by the SEA team.

Stage 4: Monitoring and evaluation

✦ Monitoring decisions taken on the PPP and the results of their implementation

It is important to monitor the extent to which environmental objectives or recommendations made in the SEA report or the PPP are being met. Information tracking systems can be used to monitor and check progress of the PPP. Monitoring of cumulative effects may be appropriate for initiatives that will initiate regional-scale change in critical natural assets. Methods and indicators for this purpose need to be developed on a case-by-case basis.

✦ Evaluation of monitoring results and feed back in PPP renewal

At some point a formal evaluation of the monitoring results should take place as part of the revision or renewal of the PPP.

Notes

1. These principles draw on a range of sources in the professional literature. In developing individual SEA methodologies and terms of reference the reader can find further information in: IAIA principles and criteria (IAIA, 2002), *www.iaia.org*; IAIA SEA Training Course Manual (Partidario, undated), *www.iaia.org*, Abaza et al. (2003).

2. Based on Sadler, 2001.

3. The Calabash Programme of the Southern African Institute for Environmental Assessment (SAIEA) seeks to increase the capacity of civil society to participate in environmental decision-making. It has developed a suite of practical public engagement tools (*e.g.* Template Terms of Reference for the Public Participation phase of an SEA) that can be used in the development of PP programmes. See *www.saiea.com* (search for Calabash).

4. Organisations such as the International Association for Public Participation have developed toolkits of communication techniques and methods and advice on how to select appropriate ones for the particular context (*www.iap2.org*; see "Toolkit").

Part II

Chapter 5. **Applications of Strategic Environmental Assessment in Development Co-operation**................................. 65
5.1. Key entry points for SEA... 66
5.2. Donor harmonisation for SEAs 69
5.3. Guidance Notes 71
A. Guidance Notes and Checklists for SEA Led by Partner Country Governments.. 72
Guidance Note and Checklist 1: National Overarching Strategies, Programmes and Plans 72
Guidance Note and Checklist 2: National Policy Reforms and Budget Support Programmes................................. 77
Guidance Note and Checklist 3: National Sectoral Policies, Plans and Programmes............................ 81
Guidance Note and Checklist 4: Infrastructure Investments Plans and Programmes............................. 86
Guidance Note and Checklist 5: National and Sub-National Spatial Development Plans and Programmes.................... 90
Guidance Note and Checklist 6: Trans-National Plans and Programmes......... 93
B. Guidance Notes and Checklists for SEA Undertaken in Relation to Donor Agencies' Own Processes 98
Guidance Note and Checklist 7: Donors' Country Assistance Strategies and Plans 98
Guidance Note and Checklist 8: Donors' Partnership Agreements with other Agencies 101
Guidance Note and Checklist 9: Donors' Sector-Specific Policies 103
Guidance Note and Checklist 10: Donor-Backed Public Private Infrastructure Support Facilities and Programmes 106
C. Guidance Notes and Checklists for SEA in Other, Related Circumstances ... 111
Guidance Note and Checklist 11: Independent Review Commissions (which have implications for donors' policies and engagement) 111
Guidance Note and Checklist 12: Major Private Sector-Led Projects and Plans..... 115
Notes 119

ISBN 92-64-02657-6
Applying Strategic Environmental Assessment
Good Practice Guidance for Development Co-operation
© OECD 2006

PART II

Chapter 5

Applications of Strategic Environmental Assessment in Development Co-operation

As noted previously, SEA can be applied in a wide range of contexts where there is a need to integrate environmental issues into decisions about policies, programmes and plans. The shift of emphasis away from projects to programme and policy development and support has created a number of particular **entry points** for the application of SEA – if development agencies and their partner governments are to achieve the same degree of environmental mainstreaming that has become standard at the project level. This chapter first introduces those "entry points" where SEA may have the most value. Next, it provides **Guidance Notes** on how SEA can be applied in the context of each of those entry points, together with checklists of the main substantive issues to be addressed and illustrative case examples.[1] In some cases, there is a growing body of practical experience in applying SEA to the entry point in question while in others SEA offers as yet unrealised and untested opportunities.

5.1. Key entry points for SEA

Almost all applications of SEA in the context of development co-operation will involve some degree of partnership between developing countries and development co-operation agencies. However, the lead roles and responsibilities in applying SEA will vary according to the activity concerned.

In the majority of cases, the management of an SEA process (including the decision on whether an SEA is required) will be the responsibility of the partner country. In this context, the main role of development agencies will be to support partner countries in conducting SEAs and to provide technical assistance and capacity-building where necessary. Table 5.1 provides an overview of the main groups of PPPs where country partners play the lead role, the corresponding lead authorities and applicable development co-operation instruments. It also indicates the potential for applying SEA to the different instruments to influence PPPs.

Development co-operation agencies have their own policies, procedures, plans and strategies to guide their operations. The may be mandated by their governing bodies or by individual governments in the case of bilateral agencies. Examples include country assistance strategies or plans, and agency-wide policies or strategies covering key issues (*e.g.* water, health). The sustainability of these strategies can be enhanced by the application of SEA. In many cases, development co-operation agencies have a specific obligation to review the environmental implications of the sector-specific policies, plans and programmes that they support. This may involve conducting SEAs of partner countries' own plans and programmes, *e.g.* the World Bank must ensure compliance with a series of environmental safeguard policies for investment lending[2] and for development policy lending. Table 5.2 lists some of the entry points and opportunities for SEA related to the internal instruments and activities of development co-operation agencies.

Table 5.1. **Key entry points for SEA: Country level**

Lead authorities	Focus area/Entry point	Instruments	Potential for SEA application
National Government and Cross-Sector Ministries (*e.g.* Departments of Finance/Planning)	National-level overarching strategies, programmes and plans	Poverty Reduction Strategy Papers National Sustainable Development Strategies 5 and 10 year plans MDG-based national development strategies	Strategic planning frameworks risk being severely flawed without SEA. There is an important opportunity to ensure environmental considerations underwrite the sustainability of strategies, plans, etc. **See Guidance Note 1.**
	National Policy reforms and Budget Support programmes	Development Policy Lending (DPL) Direct Budgetary Support (DBS) Aid-funded debt relief	Policy reforms and budget allocations create complex and often indirect environmental effects and opportunities. More up-streamed SEAs focused on institutions and governance systems for dealing with complex indirect and cumulative effects are required. **See Guidance Note 2.**
Sector or Line Ministries (*e.g.* Mining, Health or Agriculture)	National sectoral policies, plans or programmes, *e.g.* energy or health sector reform	Sector-Wide Approach (SWAp) Sector Budget Support Sector Policy Lending	The sustainability of these initiatives can be enhanced by the application of SEA, encouraging inter sector consultation and focusing on institutions and governance. **See Guidance Note 3.**
	Infrastructure investments plans and programmes	Loans Equity investment Grants Investment lending Technical assistance etc.	The scale, nature and regional or sectoral reach of major infrastructure investments require more than traditional EIA and CIA to account for induced downstream economic and social changes which may cause significant environmental and social effects, and for strategic choices available to enhance development impact. **See Guidance Note 4.**
Sub-national, regional and local Governments	National and sub-national spatial development plans and programmes	Technical assistance and investment	The sustainability of initiatives can be enhanced by the application of SEA, encouraging multi-stakeholder consultation. **See Guidance Note 5.**
International/ transboundary agencies	Trans-national Plans and programmes (including multi country plans and investment programmes)	Technical assistance and investment	SEA and CIA can increase the sustainability of these programmes, and also reduce the risk of conflict by encouraging international co-operation. **See Guidance Note 6.**

Table 5.2. **Key entry points for SEA: Development agencies' own activities**

Lead authorities	Focus area/Entry point	Instrument (partner or donor)	Potential for SEA application
International (multilateral and bilateral) development agencies	Country assistance strategies and plans Development assistance frameworks	Country or region-specific assistance programmes	The sustainability of these initiatives can be enhanced by the application of SEA. Policy recommendations from Country Environmental Analysis can trigger SEA. **See Guidance Note 7.**
	Donors' Partnership agreements with other agencies	Bilateral support to intermediate agencies in aid delivery	**See Guidance Note 8.**
	Donor sector-specific policies and strategies	Policies focusing on specific priority issues, *e.g.* water and sanitation, agricultural development	An SEA requirement may exist under international obligations, national legislation or directives or an agency's own procedures. **See Guidance Note 9.**
	Donor-backed public private infrastructure support facilities and programmes	Public/Private sector support for infrastructure programmes	**See Guidance Note 10.**

To sum up, an SEA can be initiated in response to:

- A legal requirement at the international or national level for SEA when developing certain PPPs.

- A decision by a ministry in a partner country concerned about an existing or proposed PPP's impacts.
- The need to meet the policy requirements of a development co-operation agency for assessing the environmental impacts of the programmes it supports.
- A decision by a donor to support SEA capacity development.

Table 5.3 compares the *key features* of SEA led by partner country governments with those SEA undertaken in relation with donor agencies own processes.

Table 5.3. **SEA led by country governments and SEA undertaken in a donor agency's own processes: Key features compared**

	SEA applied to donor agency processes	SEA led by country governments
How is it initiated?	By administrative or policy requirement or initiative of environmental specialist, country or strategy manager.	Due to administrative or legal requirement in country, or request from donor agency or initiative on part of government champion.
Who does it?	Team leader and/or environment specialist.	Client country government and/or third party commissioned by the government (lead role in implementing SEA). The donor agency itself and/or commissioned third parties (may have a review function or support aspects of SEA, *e.g.* workshops, analysis as SEA input.)
Objectives	To upstream and mainstream environmental considerations in strategic decision making in order to identify opportunities and manage constraints to effective development processes.	
Measures of success	Environment issues are integrated within donor support.	Environmental issues are integrated within PPP/strategy/legislation. Environmental indicators are identified for monitoring and as an input to future changes in PPPs/strategy/legislation.
Level of effort and costs	Staff time varies based on length of donor agency process.	Varies due to length of process and complexity of design: a few thousand dollars to US$2 million. Comprehensive SEAs typically average around US$200 000-300 000.
Process/steps/ Inputs	Identify opportunities to bring environmental information to decision making during formal donor process for approval of donor support (*e.g.* review meeting of proposed programme to be supported/policy reform loan).	Identify opportunities to bring environmental information to decision making during formulation and implementation of PPP/strategy/legislation (*e.g.* Cabinet level discussion of draft policy or Planning Ministry meeting to discuss multi-year sector plans).
	Identify and analyse relevant environmental issues, corresponding positive opportunities and negative aspects, corresponding institutional aspects and recommendations/suggestions as input to donor supported programme or PPP/strategy/legislation. This involves both collecting and analysing available information as well as identifying gaps in information.	
	Involve stakeholders as appropriate. Seek information and/or feedback from government, own agency sector specialists, development agencies, civil society/private sector.	Involve stakeholders and help strengthen environmental constituencies as appropriate through process.
	Bring information to the table during appropriate windows in donor approval process.	Put in place feedback mechanisms for those most affected by environmental degradation, but who do not have a strong voice.
	Identify indicators for measuring progress and identify accountabilities as part of donor agency's or country's monitoring system.	
		Allocate budget to fulfil assigned responsibilities
	Review of final product/PPP/strategy/legislation to determine level of integration of environmental recommendations.	
	Monitoring environmental outcomes in similar types of products (*e.g.* PRSCs, country programmes) over longer term to improve future donor support with respect to products.	Monitor environmental outcomes over longer term to provide feedback to future revisions of PPP/strategy/legislation and disseminate the monitoring outcomes.

SEA approaches are also relevant in a variety of other circumstances. These include, in particular, independent reviews of development agencies' involvement in specific areas or sectors and donor-supported private sector initiatives. Development agencies are involved but they are often not the drivers of these processes. The Extractive Industries Review and the World Commission on Dams are recent examples of the former. They considered the merits of the World Bank's continued investments in these respective sectors, and the compatibility with the World Bank's poverty reduction and sustainable development goals. This type of independent multi-stakeholder scrutiny and review can be viewed as a form of SEA. Private-public collaborations with regard to investments in the oil and gas sectors provide examples of the latter, where private sector organisations have applied SEA-like approaches to their investment strategies. Table 5.4 points to entry points for SEA that relate to these circumstances.

Table 5.4. **Key entry points for SEA: Related entry points**

Initiating authorities	Focus area/Entry point	Instrument	Potential for SEA application
Independent Review Commissions	Review of development agency's engagement with specific sectors (*e.g.* Extractive Industries Sector Review, World Commission on Dams)	Not an aid instrument but a review of the development effectiveness of an international agency's own engagement in a sector.	These Reviews could be considered SEAs in their own right in that they incorporate the procedural criteria for an SEA approach. **See Guidance Note 11.**
Private Sector Entities	Development or review of corporate policies, strategies and investments	Private sector strategies and investments have many points of interaction with development agencies.	SEA could contribute principles that will ensure greater rigour to corporate scenario planning, and examine risks and opportunities associated with major investments. **See Guidance Note 12.**

The Checklist below outlines *generic questions* that should be part of any SEA.

5.2. Donor harmonisation for SEAs

SEA undertaken at the macro-economic or sector-wide level may imply a significant investment of resources in terms of time and expertise, particularly from the partner country. Hence, it is important for development agencies to observe good harmonisation practices in order to alleviate the burden of administrative and other work for the relevant stakeholders likely to be involved.

Therefore, one of the first steps of an SEA is to identify:

- which other development agencies are active in the same country/sector/region; and
- whether any other development agencies have recently already carried out an SEA (or similar impact assessment) about the same or a similar plan, programme or policy or intend to do so in a near future.

In such cases, it will be prudent, if not essential, to consider undertaking an SEA as a joint exercise by two or more development agencies. Different situations may be envisaged:

- Delegate the SEA so that it is led by the development agencies acknowledged to have the best experience in the area and/or country. Other development agencies would then be expected to contribute financially to support their share of the cost.
- Carry out an SEA with the host country on behalf of other development agencies (the inverse situation of the previous one).
- Combine staff and financial resources to undertake a common, multi-donor SEA process.

Generic checklist: Questions for all SEAs

Principles and scope

- Have adequate principles, criteria and indicators been defined for the SEA?
- Has the spatial and temporal scope of the SEA been adequately defined?
- Is there a need/opportunity for donor co-ordination in the conduct of the SEA?
- Have alternatives (to the proposed PPP) been identified and considered?

Linkage to other strategies, policies and plans

- Have all relevant strategies, policies and plans – at national to local levels – been reviewed (*e.g.* PRS, MDG-based strategy, district plan) and is the assessed PPP supportive of and consistent with their goals? Have any conflicts been taken into account in the design of the proposal?

Effects

- Have the potential direct, indirect and cumulative negative and/or positive effects (short-, medium- and long-term; environmental and social) of the proposed PPP been predicted and analysed?
- Have relevant, specific measures been identified and included to counteract/mitigate these? Alternatively, is it made clear how other national policies/programmes are mitigating the potential negative effects?
- Is there potential for enhancing positive effects? Have these opportunities been maximised?
- Has the quality of the assessment been independently reviewed?

Stakeholder engagement

- Have all relevant stakeholders had an opportunity to engage in the SEA process and to identify potential impacts and management measures?
- In particular, have the views of civil society, particularly affected communities, being included? What has been their influence in the development of the proposed PPP?

Capacity

- Is there sufficient capacity within institutions and agencies, at national and sub-national levels, to implement the specific PPP (*e.g.* to enable them to apply an environmental management framework for sub-elements); and to manage, regulate and be accountable for use of natural resources? How can these institutions be strengthened?
- Is there an institutional framework to manage environmental impacts and major environmental resource policy and institutional failures?
- Is the environmental policy framework and legislative authority in place to respond to any problems that might arise?

Influence of SEA

- Are there specific points in the process to develop the PPP where the SEA can have influence over decisions or design?

Data, information and monitoring

- Are there significant data and information deficiencies and gaps? How can these be filled?
- Are measures proposed for monitoring? Are these clear, practicable and linked to the indicators and objectives used in the SEA? Are responsibilities clear?

In parallel to this concern for a harmonised SEA approach, it is crucial to identify existing assessment systems in the host country. The SEA process should be integrated within these systems and help strengthen them where necessary. SEA processes should be applied by development agencies in a way that should not substitute for strengthening country capacity to carry out SEAs.

5.3. Guidance Notes

The following Guidance Notes give an introduction to the fundamental characteristics of SEA and a framework (with illustrations) of how an SEA could be undertaken in different circumstances. They are not intended for mechanistic application. Some of these opportunities to apply SEA have been successfully tried and tested; others, although promising, are still just possibilities. While the SEA principles apply generically to all forms of SEA, different approaches will be required in each situation. The Guidance Notes reflect the different style of SEA approach appropriate for different levels of the decision-making hierarchy. They have been grouped into categories for convenience. Inevitably, however, there is overlap between these categories.

Each Guidance Note includes:

- An introduction to the entry point.
- An explanation of the rationale for applying SEA to the entry point.
- A checklist of questions that would shape the methodology of an SEA for this type of entry point.
- Case examples designed to illustrate how SEA can be applied in practice.

A. Guidance Notes and Checklists for SEA Led by Partner Country Governments

The following Guidance Notes and Checklists correspond to the entry points outlined in Table 5.1 above, for which the developing country partners play the lead role.

1. National overarching strategies, programmes and plans
2. National policy reforms and budget support programmes
3. National sectoral policies, plans and programmes
4. Infrastructure investments plans and programmes
5. National and sub-national spatial development plans and programmes
6. Trans-national plans and programmes

Guidance Note and Checklist 1: National Overarching Strategies, Programmes and Plans

Description of entry point

Most countries have overarching national development strategies outlining long-term development objectives. These include, for example, National Sustainable Development Strategies, Five or Ten-year Development Plans, etc. In recent years, many developing countries have developed Poverty Reduction Strategies (PRSs) reflecting the need to eradicate extreme poverty and hunger as indicated in the primary Millennium Development Goal. Poverty Reduction Strategies Papers (PRSPs) describe a country's macroeconomic, structural and social policies and programmes to promote growth and reduce poverty, as well as associated external financing needs, usually on a three years time horizon.

Developing country governments have followed participatory processes involving civil society and development partners to prepare PRSs. These Strategies have become a key focus for donor assistance and are the main strategic framework through which across-the-board, pro-poor policies are developed to alleviate hunger, reduce child mortality and provide basic infrastructure. Several countries are expanding their PRSPs into national strategic documents (*e.g.* Ghana and Tanzania). Some Asian countries have begun to rationalise their Five Year Plans with the PRS approach (*e.g.* Viet Nam).

To date, Poverty and Social Impact Assessment (PSIA) has tended to be the main impact analysis approach used to inform PRSP development. But it does not cover environmental issues. Largely as a result, PRSPs completed to date have paid only weak attention to environmental concerns, as shown by World Bank analysis.[3] This is surprising given the many linkages between environment, poverty reduction and economic growth noted at the 2002 World Summit on Sustainable Development and which underwrite the Johannesburg Plan of Implementation.[4] SEA can be applied to PRSPs to determine whether poverty-environment issues have been well addressed and if the agreed programmes are environmentally sustainable. Preferably, SEA should be undertaken in an integrated manner during PRSP preparation, but it is also commonly applied to assess a completed PRSP or during the revision process.

Rationale: Linkages between poverty reduction, growth and environment

In the last few years, Poverty Reduction Strategies have become the most prominent strategic planning processes in development co-operation, and a main framework within which to address sustainable development. Such an integrated approach has been identified as necessary to achieve poverty reduction in view of the crucial links between growth, environmental degradation and poverty.[5]

- Environmental degradation hits the poorest the hardest, since poor people are directly dependent on a wide range of natural resources and ecosystem services for their survival.
- Environmental commons (such as grazing lands, waters and forests) contribute significantly to poor people's income but are vulnerable to unsustainable use.
- The poor (particularly women and children) are heavily affected by environmental health problems such as lack of safe water and sanitation, indoor air pollution and exposure to chemicals and vector-borne diseases.
- The majority of the rural and urban poor live in ecologically fragile areas and/or environments with high exposure to environmental hazards.
- Women are often more vulnerable than men to environmental degradation and resource scarcity. They typically have weaker and insecure rights to the resources they manage (especially land), and spend longer hours on collection (of water, firewood, etc).
- High economic growth is necessary in many low-income countries to achieve the Millennium Development Goals. Ignoring the environmental sustainability of growth may lead to short-run economic gains for some, but risk undermining long-term growth and poverty reduction.
- The poor's access and entitlements to natural resources are crucial to the fulfilment of basic human rights such as food, housing and health. Involving poor people themselves – and building on their views and knowledge – is thus a key to ensuring good governance of environmental resources.

Checklist: Key questions for SEA of Poverty Reduction Strategies (PRS)

Generic questions as well as decisions/activities

- Does the PRS adequately address the contribution of environmental issues to poverty reduction? Does it promote good environmental management as a means of tackling poverty reduction and economic growth?
- Does the PRS formulation team include the range of relevant competences to adequately assess environmental sustainability consequences of the PRS?
- How well are environmental concerns, from global and national to local level, incorporated into the PRS analysis and structure? How far are they integrated across all PRS goals and objectives? What links, if any, are made between environmental issues and other sectoral objectives such as poverty reduction?

> **Checklist: Key questions for SEA of Poverty Reduction Strategies (PRS)** *(cont.)*
>
> - Are sectors asked to contribute to an assessment of how the environmental and natural resources are affected by their proposed plans? Are plans modified with respect to the outcome of the assessment?
> - Has there been a review of environmental expenditure?
> - Are poverty-environment linkages used to prioritise poverty reduction sectoral targets and implementation measures?
> - What can be learned for future revisions of the PRSP document and other PRSP processes, and what are the key next steps to improve pro-poor environmental outcomes through the PRS?
> - How are the PRS objectives aligned with the Millennium Development Goals? Is the PRS consistent with or linked to the MDG-based development strategy?
>
> **Linkages/impacts**
>
> - Is there a clear understanding of the poverty-environment links within the country?
> - How much do the country's main natural resource sectors contribute to economic growth? Are there opportunities for them to be better utilised to enhance pro-poor growth?
> - Are the country's growth targets vulnerable to environment-related shocks? What needs to be carried out to improve the situation?
> - What are the levels of dependence of the poor on environmental goods and services? How much employment or income-earning opportunities do natural resources provide, particularly to the poorest?
> - Is there recognition of the effects of environmental hazards on health, livelihoods and vulnerability?
> - Are the issues of governance (including those related to illegal resource use and corruption) within the natural resource sectors openly debated? How are they being tackled?
>
> **Institutional/implementation**
>
> - Are key poverty-environment indicators included in the PRS monitoring plan? What is the institutional capacity for carrying out poverty and environmental monitoring and evaluation? What indicators have been developed for monitoring?
> - What steps are in place to follow through to budget allocations and implementation programmes – taking account of priorities based on an analysis of poverty-environment linkages?
> - Are any natural resource/environmental agencies undertaking new or expanded activities or changing the way they work to implement PRS related activities?
> - Are financial resources sufficient to implement the activities identified as needed to ensure sustainability, including law enforcement? Have the needed resources been channelled down to regional and local levels?
> - Is co-ordination across government sufficient to deliver on the cross-cutting environmental and natural resource issues over time? Have new cross-departmental committees, groups and/or new units inside other ministries been established to deal with environmental issues? What changes to the institutional make-up are still needed?

Case examples

> ### Case example 5.1. Mainstreaming environment into Poverty Reduction Strategies – SEA of Poverty Reduction Strategy Papers: Uganda and Rwanda
>
> **Background and objectives**
>
> Several bilateral donor agencies have supported initiatives to mainstream' environment and sustainability issues into PRS papers during the poverty reduction strategy formulation process in Rwanda and Uganda. These initiatives started when the draft PRS papers were close to completion. This was not an ideal situation but illustrative of the problems faced in undertaking SEAs in practice. In this situation, the first task was to review the PRS papers and then make amendments to adequately address the environmental threats and opportunities that had been overlooked. Both draft PRSPs had focussed heavily on growth of the agricultural sector and, in the process, had overlooked a range of important issues with environmental implications.
>
> **Approach and outcomes**
>
> - Good links were established with the PRSP drafting teams.
> - Close engagement with the national environment agencies was achieved.
> - Engagement with sectoral ministries was established.
> - Parallel provision was made for capacity-building in environmental management.
> - Commitment was made to a wide variety of follow-up actions.
>
> In both countries, engagement with champions' within both governments and donors was important. This is particularly important in ministries that are traditionally considered less influential parts of government and where a lack of general political and administrative will/support may need to be addressed. The **Rwanda** case clearly demonstrated the benefit of national champions having real ownership.
>
> The **Uganda** case also demonstrated the need for and benefits of follow-up actions to maintain sustainability during the implementation of the PRSP. These included the following activities:
>
> - Provision of environmental input to national budgeting processes.
> - Insertion of poverty/environment benchmarks in the PRS Credit.
> - Environmental integration with lead agencies' of government.
> - National review of environmental governance.
> - Technical assistance – strategic placement of environmental advisers in ministries/agencies.
> - Monitoring poverty/environment indicators, in co-operation with government.
> - Preparation of poverty/environment planning guidelines for local authorities and other critical government arms.

> ### Case example 5.2. **Incorporating environmental considerations into Ghana's Poverty Reduction Strategy processes: SEA of Poverty Reduction Process**
>
> **Background and objectives**
>
> Ghana's Poverty Reduction Strategy (GPRS), published in February 2002, identified environmental degradation as a contributory cause of poverty. However, overall, the GPRS treated the environment as a sectoral or "add on" matter rather than as a crosscutting issue. This presented major problems as many of the policies relied on utilisation of the country's rich natural resources whose future yield was threatened by significant negative environmental impacts resulting from implementation of the policies themselves.
>
> Ghana's Government decided to carry out an SEA so that environmental issues could be mainstreamed in a revised GPRS. The SEA aimed to assess the environmental risks and opportunities represented by the policies encompassed by the GPRS, and to identify appropriate management/mitigation measures to ensure that sound environmental management contributed towards pro-poor sustainable growth and poverty reduction in Ghana.
>
> **Approach**
>
> The SEA was led by the National Development Planning Commission and Environmental Protection Agency (EPA) and undertaken in collaboration with the Netherlands Embassy in Accra with technical advice from the UK Department of Foreign Investment (DFID) and the Netherlands Commission for EIA. The full SEA commenced in May 2003 and comprised two distinct elements: a top-down assessment of the impact of the policies contributed by 23 ministries to the GPRS and a bottom-up exploration of the issues raised by implementation of policies at district and regional levels. The SEA focused on:
>
> - Review of the extent to which environmental opportunities and risks were recognised and addressed under the five linked GPRS themes of macro-economy, production and gainful employment, human resource development, the vulnerable and excluded and governance.
> - Detailed analysis and discussion on each policy leading to recommendations for revision, replacement and addition.
> - Examination of the sustainability of district level plans – the principal vehicles for implementing the GPRS.
>
> **Outcomes**
>
> All the key ministries were exposed to SEA processes and guided on how to incorporate environment in policy formulation. Benefits of SEA included refinements to development policy, alterations of district level plans and revision to planning guidelines to include environmental considerations in planning at Sector and District levels. National planning guidelines are now formally required as part of policy formulation and budgeting in the GPRS process. Active participation of stakeholders (including politicians, the finance sector and NGOs) and use of SEA at all levels of decision making has led to greater emphasis on the role of SEA in improving the processes whereby the policies themselves are translated into budgets, programmes and activities. This harmonised development objectives, including alignment with the MDGs and other regional and national strategies. SEA also changed of attitudes of officials responsible for planning and budgeting, seeking win-win opportunities in integrating environment in PPPs. The 2006-09 GPRS is now being drafted with direct inputs from the SEA team.
>
> *Source:* Jean-Paul Penrose, DFID (pers. comm.), Netherlands CEIA (2003) and Peter Nelson (2003 and pers. comm.). Christine Okae Asare (pers. comm.).

Guidance Note and Checklist 2:
National Policy Reforms and Budget Support Programmes

Description of entry point

Financial support is increasingly provided by development agencies to national-level policies and to government budgets via, for example, Development Policy Lending (DPL)[6] and Direct Budget Support (DBS) programmes. New mechanisms are also emerging that aim to increase and enhance the availability of finance for national-level development to achieve the MDGs in poor countries, *e.g.* the International Finance Facility. Their aim is to assist the borrower through a programme of policy and institutional reforms and implementation that promotes growth and achieves sustainable reductions in poverty. The specifics of this would be determined after a review of the country's policy and institutional framework. Institutional capacity would also be reviewed to determine the country's ability to implement effectively the programme to be supported. This would ensure that the two are consistent and supportive of each other.

Budget support, for example, can often be seen as a means to help a country partner government to translate its Poverty Reduction Strategy into medium-term expenditure frameworks and annual budgets (see Guidance Note 1). Partner governments and donors usually make an annual joint review of the country's development, and this provides an opportunity to also evaluate environmental achievements in relation to the PRS. This review provides a forum for dialogue with potential to influence the budget and further reform programmes. The support provided will then often include a combination of general budget support and policy and institutional actions (including economy-wide reforms such as tax reforms, privatisation, decentralisation, trade liberalisation and sectoral reforms). Lending to support specific policy reforms can result from an assessment of the country's policy and institutional framework, including the country's economic situation, governance, environmental/natural resource management, and poverty and social aspects. Assistance at this level increasingly takes the form of "pooled support" whereby several development agencies collaborate to support national-level budgets.

National-level policy reform and the resulting budget allocations can have complex effects, including profound environmental implications, positive or negative. Many of these impacts will be indirect (or "second round"). They can materialise over the short-, medium- and long-term, in sectors not concerned by the reform process itself, including notably the environment. Assessing these effects is an important part of an SEA, and includes identifying the sectors potentially affected, estimating the possible new pressures placed on the institutions and agencies concerned, and laying the basis for possible compensatory measures.

Indirect impacts can be traced by reference to the "transmission channels" of reform and their implications for different groups. These include:

- *Prices* – which determine real purchasing power through direct effects on consumption (*e.g.* if households pay more for petrol) and indirectly through effects on demand for substitute products (*e.g.* fuel-wood). Fuel price increases may intensify pressures on forests if fuel-wood is a ready petrol substitute.

- *Employment* (informal or formal) – it provides the main source of income. Some policies, such as the thorough restructuring of an industrial sector may, for example, shift demand for labour leading to a surge in unemployment in given geographical areas.

- *Assets/values* (financial, physical, natural, human or social) – *e.g.* reduction in traffic congestion and air pollution may increase the value of housing and land in impacted areas.

Examples of linkages between macro-level policies and the environment include.[7]

- The prices of key inputs, such as electricity and water, condition the economic feasibility of water and energy saving measures by consumers or producers.
- Shift in exchange rates influence the costs and benefits of importing inputs (*e.g.* fertilisers) and capital goods (*e.g.* modern, energy efficient industrial equipment) with positive or negative impacts on the environment.
- The balance between various forms of taxation (income taxes, sales taxes, levies on natural resource extraction, labour taxes, export taxes etc.) will influence the economic feasibility of alternative productive activities and the choice between different inputs and outputs.
- Legal provisions relating to land tenure, property rights, and access rights to natural resources influence the profitability of natural resource-based activities (*e.g.* logging, agriculture, and fisheries), supporting or discouraging sustainable management practices.
- The structure of import tariffs and registration fees on vehicles may impact polluting emissions from transport (in many African countries, for example, imports of second-hand minibuses with high emission levels are encouraged through favourable fiscal treatment).
- Unless proper attention is given to performance control and tariff structures, the commercialisation or privatisation of delivery of public services such as waste management and water provision may lead to adverse environmental and human health impacts.

Rationale: applying SEA to national-level policies and budget support

Currently, SEA is the main existing approach to integrate environmental considerations into national-level policies and programmes. An SEA will look at the environmental implications of the reforms and help to ensure that the environmental consequences of policies, plans or programs are identified before adoption, that feasible alternatives are properly considered and that relevant stakeholders are fully involved in the decision process. Although an SEA can result in modification of proposed policies, experience shows that SEA has generally lead to the identification of mitigation and complementary policies to minimise negative impacts or maximise positive ones.

Checklist: Key questions for SEA of policy reform and budget support

Generic questions as well as decisions/activities

- What is the nature and scope of the Development Policy Lending and Direct Budget Support programme and hence what level of analysis is to be applied?
- Does the DPL or DBS explicitly support national environmental and sustainable development policies?
- Is there sufficient quantitative (as opposed to solely qualitative) information for the SEA? Has there been a review of the PRS? Have other institutions evaluated the PRS and what was the outcome as it relates to policy reform and budget support?

> **Checklist: Key questions for SEA of policy reform and budget support** *(cont.)*
>
> **Linkages/impacts**
>
> - Have the linkages between the proposed reforms and the environment been identified? Are these well understood or do they need further analysis?
> - Have the "transmission channels" been identified and indirect or unexpected environmental effects traced?
> - Have specific measures been devised to counteract the potential negative effects of the proposed reforms? Alternatively, are there clear linkages to identify how other national policies/programmes are mitigating the potential negative effects?
> - Is there potential for enhancing positive effects? Have these been maximised?
> - When will effects materialise – within the life of the operation or beyond?
> - What is the environment's capacity to adjust to the direct and indirect effects of the DBS programme?
>
> **Institutional/implementation**
>
> - What is the proposed delivery method for the DPL or DBS programme?
> - Are there any market, policy or institutional failures or other forms of distortions in other parts of the economy that need to be taken into account in the operational design of the DPL or DBS programme?
> - What is the institutional capacity to understand and respond to the linkages between policies and the environment on a sustained and adaptive basis? What steps can be taken to support this process?
> - What is the level of public concern at the local, national and international levels about the DBS programme? Does the government require the public disclosure of information and data about the environment and natural resources? Is public engagement part of the policy process?

Case examples

> **Case example 5.3. SEA of poverty reduction credit, Tanzania (budget support)**
>
> **Background and objectives**
>
> In 2004, Tanzania developed its second poverty reduction strategy: the *National Strategy for Growth and Reduction of Poverty* (NSGRP). It followed an extensive consultative review that built on the outputs from the national poverty monitoring system, and involved a wide range of stakeholders down to the village level. The NSGRP was strongly outcome-focused with increased attention to growth and governance. It set out to incorporate key cross-cutting issues, including the environment, as integral to the strategy and not as an after-thought.
>
> Tanzania has benefited from increasing levels of direct budget support with the World Bank's contribution through a poverty reduction strategy credit (PRSC). In 2004, Tanzania undertook an SEA of the second PRSC (PRSC 2) to assess the cumulative environmental and socio-economic impacts of PPPs supported by the PRSC 2, and to suggest appropriate mitigation and monitoring as well as additional capacity strengthening measures.

Case example 5.3. **SEA of poverty reduction credit, Tanzania (budget support)** *(cont.)*

Approach

The first challenge was to determine which PPPs the SEA should focus on since, technically, PRSC funds could be used to support the implementation of any Government PPP. It was decided to concentrate on programmes highlighted as "triggers" for the release of PRSC funds. These were detailed in the Performance Action Framework, which identifies key actions agreed between the Government, the World Bank and other development partners providing budget support. It was agreed that the SEA would address reform of the Crop Boards, the development of a strategic plan to operationalise implementation of the Land Act and Village Land Act, introduction of a Road Act to support the maintenance and rehabilitation of district roads, and a new business licensing system. The SEA would also consider the capacity for environmental management and assessment.

Outcomes

The study identified potential impacts and mitigation measures regarding the adverse effects of the reforms on the environment. It found that whilst Tanzania has many PPPs that could reduce identified environmental impacts, the main obstacle is limited capacity to implement, enforce and monitor plans and programmes.

Source: Eric Mugurusi, Division of Environment, Tanzania, and David Howlett (UNDP, Tanzania).

Case example 5.4. **SEA for policy reform in the water and sanitation sectors in Colombia (development policy lending)**

Background and objectives

The Colombian Ministry of Development led this SEA, with consultant technical assistance. An interdisciplinary group consisting of specialists from the Department of National Planning, the Ministry of Environment, and the Ministry of Development, prepared the terms of reference with assistance from the World Bank task team. It requested the consultants to quantify various negative externalities: deterioration of water quality, inefficient water use, and impacts associated with the construction and maintenance of public works.

Approach

The SEA was integrated early in the process of designing sector reforms and the SEA consultants conferred with relevant authorities, particularly the Ministry of Economic Development and the Ministry of Environment. An inter-agency committee was formed to facilitate joint work on environmental management issues in the water and sanitation sector. Public engagement played a prominent and early role in this SEA with two nationwide workshops. In addition, the Ministry of Economic Development and the Ministry of Environment consulted with a number of government agencies, NGOs, private utility operators, and professional associations, among others.

Outcomes

The SEA:

- Developed many recommendations to reform aspects of both the effluent charge system and the waste water discharge standards used in Colombia. These recommendations were intended to attract private investment to the water sector by reducing uncertainty associated with environmental controls.

> **Case example 5.4. SEA for policy reform in the water and sanitation sectors in Colombia (development policy lending)** *(cont.)*
>
> - Recommended changes to reform Colombia's EIA regulations.
> - Defined the environmental obligations of local authorities and utility operators.
> - Made recommendations for capacity-building and institutional strengthening for environmental management.
>
> The Ministry of Economic Development and the Ministry of Environment agreed on strategies for jointly implementing some of the SEAs recommendations in the final agreement with the World Bank.
>
> *Source:* World Bank (in press).

Guidance Note and Checklist 3: National Sectoral Policies, Plans and Programmes

Description of entry point

Support is often provided by development agencies to an entire sector. For example, this may include support for the privatisation of the energy sector in a given country or for assistance to the education sector. In many cases, several development agencies collaborate to support a sector investment programme and accompanying policy and/or institutional reforms agreed with the government. This is sometimes referred to as a Sector Wide Approach (SWAp).

Sector programmes cannot be viewed in isolation, as there are clear linkages between sectors. The health sector, for example, has linkages with virtually all other sectors, such as water, agriculture and energy. SEA should take explicit account of the opportunities for co-ordinating or integrating programming with other sectors, as well as identify the cumulative impacts arising from these sector programmes. Opportunities will also be identified for further development of sector-based issues within national policy and strategy, whilst highlighting any capacity needs.

Rationale: SEA and national sectoral policies, plans and programmes

Development agencies support country-led sector reforms, investment programmes and technical assistance – either through lending, budget support or programme grants. These cover a wide range of sectors, most of which have significant environmental issues, for example agriculture, health, transport, energy, urban development, land use, forestry, mining, industry, fisheries, water and tourism.

The benefits of applying SEA to national sectoral policies, plans and programmes are well recognised by development agencies and governments. The scale and nature of programmes require more than traditional EIA. SEA examines the environmental risks associated with the reforms, support and investments in a sector, and enables a framework for environmental management and monitoring to be agreed and built into the specific elements of the sector programme and its implementation mechanisms. SEA can also influence the overall shape and design of the sector programme by focusing on the linkages of the sector in question with other sectors and the possible cumulative environmental effects of the current programme.

The tools used to apply SEA are mainly impact-focused, but the outcomes are likely to be focused on the mechanisms and institutional capacity to address specific environmental risks as the programme is implemented.

> ### Checklist: Key questions for national sectoral policies, plans and programmes
>
> **Generic questions as well as decisions/activities**
>
> - What are the objectives and proposed key mechanisms of the national sectoral policy/programme?
> - Have the main policy instruments that steer the development of the sector been clearly identified?
> - What are the main environmental and social impacts and risks traditionally associated with this sector? Are these linked to issues addressed by international instruments (*e.g.* the Conventions on climate change, desertification and biodiversity)?
> - Is this sector a priority issue in national-level policies and strategies? If not, does the sector programme contribute to the development and integration of sector-based issues within national policy and strategy?
> - How does this sector programme contribute to sustainable development objectives within national policy and strategy?
> - What are the alternatives to the sector programme elements under consideration?
> - What are the key decision points in designing, consulting with relevant stakeholders, and agreeing the national sectoral programme? Are there any environmental checkpoints? What kind of environmental analyses are required for approval at such key decision points?
>
> **Linkages/impacts**
>
> - What are the key linkages between the sector programme elements and the environment?
> - How might social, environmental and natural resource issues in the sector influence favourably, or affect, national development priorities?
> - Have environmental and social direct and indirect effects and opportunities been considered in the sector programme? How will these effects and opportunities be managed and implemented?
> - Who are the relevant stakeholders for the sector programme? Are their priorities and environmental concerns well understood in relation to the sector programme?
> - Are there any potential areas of conflict, either within the sector or with other sector programmes?
> - Has there been a review of environmental expenditure?
> - Is institutional capacity within and outside the sector able to deal with, adapt to, and take advantage of, the environmental and social effects and opportunities that may arise because of the sector programme?
> - Does the sector programme involve dialogue and co-ordination with other ministries for cross-sectoral interventions?

> **Checklist: Key questions for national sectoral policies,
> plans and programmes** *(cont.)*
>
> **Institutional/implementation**
>
> - Is there a need for donor or sector ministry co-ordination in the conduct of the SEA?
> - Have the use of participatory methods and processes been formally adopted? Have weak and vulnerable stakeholders made their voices heard? Are communities involved in decision-making?
> - Have appropriate indicators for monitoring and development been included within the sector programme?
> - Is there capacity in the relevant ministries to integrate sector-specific issues into national strategy?
> - Is there capacity for planning for conflict resolution and mediation?
> - Does the sector programme promote education and awareness raising?

Case examples

> **Case example 5.5. The Kenya Education Support Programme**
>
> **Background and objectives**
>
> The Kenya Education Support Programme (KESSP) is the programme through which the Government of Kenya, development partners, civil society, communities, and the private sector have come together to support education sector development for the period 2005-10. The programme fits within the framework of national policy set out in the Economic Recovery Strategy (ERS) and has been developed through a Sector Wide Approach to Planning (SWAP). The donor community views initiatives such as KESSP as central to the achievement of international development objectives such as the Millennium Development Goals. An SEA of KESSP was undertaken before the investment programmes had been fully designed, so it was important that the SEA i) assessed the potential impacts that the investment programmes might have in tackling some of the key (and crosscutting) social and environmental issues related to education; and ii) provide guidance on how to mitigate these potential impacts in the design and implementation of KESSP's investment programmes.
>
> **Approach**
>
> The SEA aimed to:
>
> - Provide an environmental and social situational analysis, by identifying key issues and stakeholders in the education sector in Kenya.
> - Through fieldwork, stakeholder interviews and desk research, identify the likely strategic environmental and social impacts of the KESSP and analyse the severity, significance and risk of those impacts.
> - Develop measures to manage or mitigate any negative impacts identified, and to enhance any positive impacts.
> - Integrate these measures into an Environmental and Social Management Plan which could be mainstreamed into the design and operation of the programme.

II.5. APPLICATIONS OF STRATEGIC ENVIRONMENTAL ASSESSMENT IN DEVELOPMENT CO-OPERATION

Case example 5.5. The Kenya Education Support Programme (cont.)

- Make recommendations for the design of the KESSP, including identifying gaps and opportunities, as well as potential cost savings.
- Make recommendations for any further studies needed.

Outcomes of the SEA:

- Influenced the design of the KESSP at early stage.
- Strengthened the environmental and social sustainability of implementation.
- Made institutional recommendations to enhance implementation.
- Improved donor co-ordination by maximising the use of resources, avoiding duplication of effort and integrating different donor aims and priorities.

Source: DFID/ERM (2005).

Case example 5.6. Sector EA of Indonesia Water Sector Adjustment Loan (WATSAL)

Background and objectives

The Government of Indonesia decided that several sectors, including the water sector, needed significant reform. They requested financial and technical support from the World Bank through a Sector Adjustment Loan. The Bank task team and counterparts in Indonesia undertook a voluntary pilot SEA for the project. The goals of the assessment were to:

- Inform the government and Bank management of the environmental risks associated with the policy reforms.
- Prescribe mitigation and monitoring procedures to soften adverse impacts.
- Ensure that people affected by the reforms were given a voice in formulating the analysis.

Approach

- Representatives from NGOs, academia, government, and the public together defined the terms of reference of the consultation process. Participants agreed on the principles and method for the assessment.
- Preparatory visits were then made to the provinces and districts, while case materials were prepared for dissemination to the stakeholders throughout subsequent rounds of consultation.
- The first round of consultation was held at three levels – province, district, and village – to inform the groups of the purpose and process of the loan and policy reforms, as well as identify possible impacts and devise measures to counter them.
- A draft sectoral EA report was then prepared.
- A second round of consultations followed with the same groups to share findings of the report, seek final recommendations, and record opposing views.
- A national meeting of representatives from government, NGOs, and public organisations was convened to discuss the WATSAL principles and the draft sectoral EA report.
- The final version of the report was produced. It included proposals for alternatives and mitigation for each policy reform item.

Case example 5.6. **Sector EA of Indonesia Water Sector Adjustment Loan (WATSAL)** *(cont.)*

Outcomes

This approach contained several elements of SEA good practice as it was:

- Focused and well-timed. The consultation process started early enough to influence final decisions while not delaying the WATSAL.
- Transparent. The study fostered dialogue and openness by including all stakeholders in the discussions and process.
- Participative. Stakeholders from all regions and levels of society and government participated.
- Influential. Many of the suggestions emerging from the consultations were included in the final design of the loan's reform agenda.

Source: World Bank (1999, 2000).

Case example 5.7. **Energy Environment Review in Iran and Egypt**

Background and objectives:

The Energy and Environment Review (EER) is a specific approach proposed in the World Bank's *Fuel for Thought: An Environmental Strategy for the Energy Sector* as an instrument to help set operational priorities for mainstreaming the environment in the context of the World Bank's policy on support to the energy sector.

The Energy Sector in Iran and Egypt was subject to a review because:

- The sector includes key natural resources (oil and natural gas reserves).
- It is a major source of pollution.
- It is an important sector for environmental mainstreaming.

Approach

An Energy and Environment Review process involves:

- Analysis of the current situation with regards to energy generation and use.
- Evaluation of the growth prospects with regards to energy generation and use.
- Identification of environmental issues induced by the generation and use of energy, and damage cost estimates.
- Evaluation of the extent of contribution to climate-change through emission of greenhouse gases.
- Evaluation of the proposed mitigating measures for the previously identified environmental problems.
- Conclusions and recommendations, and a proposal for an action plan.

Outcomes of the EER in Iran

- Increase in the price of energy fuels (gasoline, diesel, and electricity should reach real market values by 2009).
- Reduction of gasoline price subsidies (the price of gasoline should reach its market value in 2009).
- Diesel fuel price was increased by 15%.

> **Case example 5.7. Energy Environment Review in Iran and Egypt** *(cont.)*
>
> - Electricity price increased by 20%.
> - Enabled the Carbon Business Finance Unit to commit to buy US$ 50 million of carbon emission reductions.
> - Helped the Bank in convincing the government to ratify the Kyoto Protocol.
>
> **Outcomes of the EER in Egypt:**
>
> - Used by the Bank for advancing its policy dialogue in the Energy Sector.
> - Enabled the Carbon Business Finance Unit to commit to buy US$ 50 million of carbon emission reductions.
> - Helped the Bank in convincing the GOE to ratify the Kyoto Protocol.
> - Improved the Country Environmental Analysis of Egypt.
> - Increased penetration of natural gas in industrial sectors as well as CNG in taxies.
> - Improved efficiency of electrical transmissions and distribution systems.
> - Increasing the price of diesel.
> - Vehicle Emission Testing (VET) in Greater Cairo.
>
> *Source:* Arif (2005).

Guidance Note and Checklist 4: Infrastructure Investments Plans and Programmes

Description of entry point

The importance of SEAs for large infrastructure investments or multi-project infrastructure programmes is clear. Impacts are highly significant, and often both regional and cross sectoral in nature. SEA enables the assessor to link policy tools with elements of the investment ensuring a full strategic integration of environmental dimensions at the very early stages. As part of the SEA, the aims, principles and priorities of the infrastructure policy will be addressed. This in turn will influence the strategy of the investment plans. A key problem faced is that the plans being reviewed tend to be highly abstract, as specific projects are not yet defined. It is important that all relevant institutions be involved in the SEA and that alternative planning and management options be compared in an integrated way. Only then will decision makers have the relevant information to enable them to make a sustainable decision.

Infrastructure investments will tend to consist of a set of related projects. As the SEA progresses, it will help to structure and focus environmental analysis on the key environmental benefits and costs of each stage, and also consider the cumulative impacts of all related developments.

Development agencies often provide support to such projects in the form of financial or technical assistance. Such support may be conditional on some form of strategic assessment being undertaken in order that the complexities associated with projects undertaken over a range of temporal and geographical scales can be adequately considered. The identification of capacity development needs and relationships with other relevant plans and programmes is particularly important.

Rationale: SEA and major infrastructure investments

Development agencies may provide support for major or multiple infrastructure projects in a number of ways including equity investment, grants, loans or technical assistance. The scale, nature and significance of this type of investment makes a strategic level assessment more appropriate than a traditional EIA and/or cumulative effects assessment, and can sometimes be a requirement of the assistance.

An important feature of an SEA for major infrastructure investments is that it identifies how the projects are affected by external factors. This involves documenting any existing environmental issues, as well as clearly identifying links with other relevant plans and programmes. In turn, this then allows potential cumulative impacts to be identified and ensures that all feasible options and alternative plans are known. In this way, the siting of the project(s) and the technologies to be adopted can be optimised.

In the case of long-lived infrastructure or networks, (*e.g.* large-scale dams, road of railroad networks) this will include assessing the likely impact of climate change, which may significantly affect important parameters (*e.g.* seasonal variabilities in water flows, temperatures, incidence of extreme weather events) within the planned useful life of the infrastructure facilities.

The SEA approach allows the planning of infrastructure projects to be integrated with land and environment planning at an early stage. This is done through a participatory process, allowing stakeholders at all levels to consider the rationale behind the identified project and discuss social, economic, land use and environmental needs/constraints. In this way, decision making is optimised.

SEA will address environmental objectives established at an international, national, district or community level that are relevant to the plan, and examine how those objectives and environmental considerations have been taken into account during preparation of the investment. It will also assess the policy environment in which the investment is to take place, and identify where capacity development is needed to help the institutional actors implement and manage the project.

Key questions for infrastructure investments, plans and programmes

Generic questions as well as decisions/activities

- Is the proposed investment programme in line with the aims, principles and priorities of the country's infrastructure and investment policy/framework?
- Are the aims and objectives of the proposed investment(s) clear?
- Are indicators and targets defined where appropriate?
- Have the local/district/provincial land use plans been reviewed and, where appropriate, been taken into account in the design of the investment (in the sector of the investment)?
- Have environmental objectives established at international, community or national level been fully considered during investment planning?
- If the investment forms part of a hierarchy of projects, has duplication of assessment been avoided?

> **Key questions for infrastructure investments, plans and programmes** *(cont.)*
>
> - Have strategic and upstream alternatives (within and beyond the sector of the investment) been analysed and compared? Has the best overall alternative been selected in a transparent, informed and rational manner? Has that alternative formed the foundation of the design of the investment?
>
> **Linkages/impacts**
>
> - Have the linkages between the proposed investments' development objectives and the environment been identified? Are these well understood or do they need further analysis?
> - Have the potential indirect and cumulative (short, medium and long term) environmental and social impacts of the investment been evaluated, have relevant mitigating measures been identified and included in the design of the investment and its companion programs?
> - Are there major risks from the investment that have potential significance beyond the immediate project area? Is the investment under risk from environmental degradation created outside the project's influence?
> - Is the infrastructure or network concerned likely to be affected by climate change? Have the key assumptions underlying project design been examined in this light, over the relevant time horizon? Is the scientific basis for making such assessments available and sufficient?
> - Have potential cross-border and transboundary effects been identified? If yes, has notification/information exchange taken place, and prior to major strategic decisions being made for the investment?
> - Are there opportunities for regional development benefits to be achieved?
>
> **Institutional/implementation**
>
> - Is there a need for institutional strengthening and capacity-building for the institutional actors with responsibility for implementing the investments; and enforcing environmental regulations, including access to the judiciary to those affected?
> - Is there a plan for the meaningful participation of weak and vulnerable stakeholders in infrastructure investment planning and specific large infrastructure projects?
> - How are the views of civil society being included? What has been their influence in investment planning?
> - Will the infrastructure investments encourage productive partnerships at local and regional level?

Case examples

> ### Case example 5.8. Regional environmental assessment of Argentina flood protection
>
> **Background and objectives**
>
> A Regional Environment Assessment (REA) was undertaken for an investment programme to protect communities occupying the flood plains of the Paraguay, Parana and Uruguay rivers in northern Argentina. This region had suffered enormous losses from periodic flooding, but this natural process also sustains ecological systems and many forms of productive activities. A "living with floods" strategy was therefore developed with proposed flood defence construction works. Non-structural measures were also introduced, including strengthening institutional capacity and co-ordination to deal with periodic flooding, and upgrading flood warning systems and technical assistance support.

> **Case example 5.8. Regional environmental assessment of Argentina flood protection** *(cont.)*
>
> **Approach**
>
> The REA was initiated at an early stage of the decision-making process and included:
>
> - Description of the interaction of hydro-ecological and socio-economic systems of the region.
> - Screening of potential investments to select sub-projects with clear economic, social and environmental benefits.
> - Analysis of alternatives for each site using criteria of least possible interference with natural flooding patterns.
> - Analysis of the cumulative effects of all flood protection projects.
> - Public consultation aimed at improving the design of all sub-projects.
> - Design changes to take into account the results of the REA and public consultations.
> - Identification of mitigation and monitoring measures.
> - Identification of institutional weaknesses in dealing with the flood problem.
> - Recommendation for a regional action plans to address the issues identified.
>
> **Outcomes**
>
> The study found that many ecosystems and human activities depend to a great extent on the periodic floods. This had a direct impact on the way the project was designed. Criteria for the selection of investments were modified to ensure that flooding would continue, but not threaten human well-being and economic infrastructure. The study also documented the extent to which wetlands, gallery forests and aquatic ecosystems of the tributaries to the three rivers were threatened by human activities. It found that the most disruptive activities were road construction, followed by poorly planned urban expansion and effluent from the meat packing industry. The REA assisted the design of four key project components to help improve the environmental and economic benefits of the project:
>
> - Strengthening EA procedures in key institutions within the seven provinces.
> - Technical assistance for urban environmental management.
> - Environmental education and awareness programmes in communities benefiting from protection works.
> - Support to protection and management initiatives for wetlands and other ecosystems.
>
> Perhaps the most important outcome of the Regional Environment Assessment was its direct contribution to screening of all potential investments under the project. It helped reduce the number of possible sub-projects from 150 to 51, all with a clear economic, social and environmental justification. Once these sub-projects had been selected, the REA team prepared project-specific EAs for each one. When they were completed, the REA team returned to examine the likely cumulative impacts of all the 51 sub-projects, to ensure that such impacts would be minimised.
>
> *Source:* World Bank (1996); Kjørven and Lindjhem (2002).

Guidance Note and Checklist 5:
National and Sub-National Spatial Development Plans and Programmes

Description of entry point

Spatial development planning is most commonly undertaken at national or regional levels and districts and has many forms. It aims to provide integrative frameworks for various economic, social, environmental interventions and is often closely linked to budget-making processes. Given their integrative nature, there are clear opportunities for undertaking SEAs.

SEA can anticipate and improve the overall environmental effects of proposed patterns of spatial development, and of multiple individual projects. They can improve foresight regarding potential effects of future plans and in the longer-term promote a more open, transparent and evidence-based planning culture.

Experiences from elaboration of spatial development plans in EU accession countries show that SEA can be effectively applied as a component of donor support for integrated planning.

Rationale: SEA and regional development programmes and plans

SEA applied to spatial/regional plans or programmes provides an important opportunity to integrate sustainable development approaches within the decision-making process. It encourages multi-stakeholder consultation and ensures that the environmental consequences of plans and programmes are identified and assessed during preparation and before their adoption. Integration of the environmental dimension at all stages (*ex ante*, interim and *ex post*) of evaluating and implementing a programme/plan enables the competent authorities to carry out changes and improvements throughout the life of the programme/plan, as appropriate. The SEA process will search for alternatives when considering the environmental, social and economic impacts of the proposal. Plans and programmes are often very different in their scope and content, so the SEA process can vary significantly.

Planning at this level is normally controlled by a legislative framework, but this depends on the degree of centralisation/decentralisation of the spatial or development planning process. Requirements for SEAs of these plans and programmes also vary, from ministerial decisions to regulations at national, regional or local levels.

Key questions for spatial development plans and programmes

Generic questions as well as decisions/activities

- Have the development plan objectives been linked with other international, national and regional policy aims?
- Have international and national environmental standards been considered and incorporated into the planning?
- Have all zones of special environmental interest and protected status within the impacted area been identified?

> **Key questions for spatial development plans and programmes** *(cont.)*
>
> **Linkages/impacts**
> - What are the priority environmental problems in the area in question? Is there a danger these problems could be exacerbated by the proposed programmes/plans?
> - Has the spatial and temporal scope of the SEA been adequately defined? Have any relevant cumulative issues been taken into account.
> - Are the proposed developments likely to be vulnerable to the impacts of climate change? Is the necessary scientific basis to assess this matter available and sufficient?
> - Has there been sufficient effort to identify environmental improvement opportunities within the programme?
> - Have relevant mitigation measures been adequately incorporated into the development and design?
>
> **Institutional/implementation**
> - Is the role of relevant environmental authorities in the planning and implementation of regional development programmes/plans well understood? Are there any capacity-building needs?
> - Are there adequate mechanisms for the results of the SEA to be reflected in the decision-making process and strategy development – *e.g.* monitoring arrangements, management and institutional issues?
> - What is the legal and administrative framework within which regional development programming/planning and environmental polices are co-ordinated, *e.g.* through land use planning and through the process of project design, approval and implementation? Is it adequate?

Case examples

> ### Case example 5.9. **The Sperrgebiet land use plan, Namibia**
>
> **Background and objectives**
>
> The Sperrgebiet is a biodiversity-rich, desert wilderness area in southwest Namibia, which also comprises a licensed diamond mining area. It has been a prohibited area since 1908. In 1994, the exclusive prospecting and mining licenses of the non-diamondiferous areas were relinquished and considerable interests arose in the area for a variety of conflicting uses. In consultation with Namdeb (the mining licence holder) and NGOs, the Government agreed that a land use plan should be formulated to ensure long-term sustainable economic and ecological potential in the fragile Sperrgebiet before it was opened up.
>
> **Approach**
>
> An SEA-type approach was used to develop the plan, involving several steps:
> - A thorough literature review with gaps filled through consultation with specialists.
> - Development of a series of sensitivity maps for various biophysical and archaeological parameters.
> - An extensive public consultation programme that included: public workshops, information leaflets and feedback forms, land use questionnaires, and a technical workshop with selected specialists.

> ### Case example 5.9. **The Sperrgebiet land use plan, Namibia** (cont.)
>
> - The establishment of a list of possible land use options for the area and their evaluation in terms of the environmental opportunities and constraints.
> - Formulation of a vision – that the entire Sperrgebiet should be declared a Protected Area. Development of a zoning plan to provide a framework to guide immediate decisions regarding land use.
> - A technical workshop including specialists to discuss and refine the draft-zoning plan.
> - A preliminary economic analysis of the main land use options.
> - Development of an administrative framework outlining the legal processes required for land proclamation, the formation of a Management Advisory Committee and definition of its role, ecotourism models, zoning, future access control and integration into the surrounding political and economic structures. For each potential land use, guidelines were prepared outlining what needs to be included in a project-specific EIA and EMP.
>
> **Outcomes**
>
> The Land Use Plan was finalised in April 2001. In April 2004, the Sperrgebiet was proclaimed a National Park. The recommendations of the Land Use Plan were accepted.
>
> *Source:* Walmsley, SAIEA, South Africa.

> ### Case example 5.10. **SEA of the Great Western Development Strategy, China**
>
> **Background and objectives**
>
> A number of regions in eastern and central China have undergone rapid economic development in the last decade, but China's western regions remain relatively poor and underdeveloped. In response, the Chinese Government's "Great Western Development" (GWD) strategy provides a strategic framework linking over 20 national policies and a range of key construction projects. The SEA of the GWD Strategy (GWD SEA) was commissioned by the State Environmental Protection Administration (SEPA). The aim was to examine environmental consequences and risks, and investigate possible modifications to specific elements of the strategy.
>
> **Approach**
>
> The SEA applied a relatively simple methodology involving co-ordinated analysis of the possible impacts associated with the implementation of the GWD strategy. This analysis focused on a broad range of environmental media and the project team used expert panels to examine both direct and indirect impacts of the strategy. They also explored alternative impact mitigation options.
>
> Sector-based studies provided an additional level of analysis and included projections of how sectors will develop in the future. The case for increasing public participation and stakeholder dialogue was briefly explored within the report, but there were no references to any formal mechanisms for public participation within the SEA process. It is therefore unclear to what extent the SEA report addresses specific concerns highlighted by some key stakeholders.

> **Case example 5.10. SEA of the Great Western Development Strategy, China** *(cont.)*
>
> **Outcomes**
>
> The interim report contains a complex matrix of direct and indirect impacts arising from activities proposed in accordance with the GWD. Nevertheless, a simple message emerges from the analysis: the environmental situation in China's western provinces is already serious, and aspects of the GWD tend to exacerbate some crucial environmental risks. Each chapter of the interim report explores a range of mitigation measures that authorities could apply to alleviate these pressures. Additional work is required to quantify the effectiveness of these measures.
>
> Arguably the ultimate test of the effectiveness of the GWD SEA study will be its capacity to influence audiences and institutions involved in the development, implementation and monitoring of the GWD Strategy. At this stage it is difficult to ascertain whether the SEA process has increased awareness and appreciation of environmental impacts associated with the GWD proposals. The breadth and scale of the GWD strategy has made it difficult to isolate specific proposals for detailed investigation. Further work on the draft report is required if it is to effectively articulate the case for a stronger focus on environmental threats and opportunities.
>
> *Source:* Haakon Vennemo and Bartlett, in Dalal-Clayton and Sadler (2005).

Guidance Note and Checklist 6:
Trans-National Plans and Programmes

Description of entry point

Trans-national plans and programmes require the consideration of a wide range of issues, such as trade and sharing of energy resources, as well as environmental issues. An example of donor involvement in such initiatives is through programme support.

SEA provides a way to integrate environmental, social and economic concerns in planning, enabling countries to lay the basis for regional co-operation. In this context, SEA is a logical approach to decision making for trans-national and regional initiatives. By identifying the interactions and cumulative effects that cross sectoral and jurisdictional lines, the SEA provides an opportunity to explore ways to mitigate negative environmental effects and enhance the positive impacts. By involving all stakeholders in this process as early and thoroughly as possible the potential for any conflict is also reduced.

Rationale: SEA and trans-national and regional planning

The nature of trans-national and regional planning has immediate strategic implications. As such, the broader view that SEA provides is vital and generates a number of direct results:

- Summary of all potential environmental and socio-economic impacts of the plan, including possible cumulative and secondary impacts.
- Recommendations on assessment criteria to evaluate the types, extent and significance of impacts that activities are likely to generate.
- Recommendations on screening criteria to identify projects likely to have transboundary impacts.
- Identification of all areas likely to be sensitive, or particularly suitable for development.

- Recommendations on any alternative activities or developments that would reduce environmental impacts.
- Recommendations on measures that will be needed to prevent, reduce and manage identified impacts. This includes assessments of their likely cost and the extent to which they could reduce the impacts to acceptable levels.
- Recommendations on data deficiencies and therefore monitoring needs.
- Recommendations on institutional and management needs required to effectively undertake development and implementation.

Some of these elements would be generated from other forms of environmental assessment, but SEA has the advantage of taking a more holistic, systematic approach, ensuring that all environmental consequences are fully included and have an enabling environment in which they can be appropriately addressed.

Key questions for trans-national plans and programmes

Generic questions as well as decisions/activities

- Have the development objectives been linked with those of other national or regional jurisdictions?
- Have different national environmental standards been discussed and adequately incorporated into the planning?
- Have the areas likely to be sensitive or particularly suitable for development been identified jointly by neighbouring countries?
- Have clear screening and assessment criteria been developed that identify which developments are likely to have transboundary implications?

Linkages/impacts

- What are the priority environmental problems in the transboundary area? Is there a danger these problems could be exacerbated by the proposed programmes/plans?
- Have discussions taken place about communicating details of the developments to neighbouring countries?
- Have neighbouring countries the opportunity to comment on or contribute to the assessment process?
- Have transboundary public engagement challenges been addressed? (See also www.unece.org/env/eia/publicpart.html.)

Institutional/implementation

- Have the neighbouring countries agreed the institutional mechanisms for communication about proposed programmes?
- Do the countries involved have similar levels of capacity for SEA type analysis?
- Are there any regional inter-governmental institutions that can support to encourage good transboundary assessment processes? If not, what might be done to encourage their development?

Case examples

> ### Case example 5.11. **Transboundary environmental assessment of the Nile basin**
>
> **Background and objectives**
>
> In 1999, the Nile riparian countries established the Nile Basin Initiative (NBI) to fight regional poverty and promote socio-economic development. Under the NBI's Shared Vision Programme (SVP), a transboundary environmental assessment (TEA) was initiated and carried out by the Nile riparians in co-operation with UNDP and the World Bank, with additional funding from the Global Environment Facility (GEF). It includes a collective synthesis of basin-wide environmental trends, threats and priorities, and outlines the elements for a long-term agenda for environmental action for the Nile Basin. The TEA aims to be both a catalyst and a valuable resource to the Nile riparians and their international partners. The main objective was to help translate existing national environmental commitments and interest into basin-wide analytical frameworks and, eventually, basin-wide actions. The emphasis was on stakeholder awareness and involvement, water and environmental management, training and education, capacity-building, information sharing and institutional development.
>
> **Approach**
>
> Priority issues to be addressed at basin-wide, national and local levels were identified and analysed. The synthesis provided the basis to formulate the elements of an Agenda for Environmental Action with complementary preventive and curative actions to address current and emerging issues in the Nile Basin. The Agenda aimed for collaborative implementation over the next decade or more in co-ordination with other development activities. Finally, the TEA outlined transboundary activities to be addressed collaboratively in the initial implementation phase of the Agenda for Environmental Action in the form of a proposed project. Two related sets of activities informed the report: broad and participatory national consultations; and a USAID scoping study for a multi-country technical background paper.
>
> Transboundary environmental threats were prioritised and these guided the formulation of a first basin-wide project for environmental action within the SVP. This Action Project has been designed to encourage more effective basin-wide stakeholder co-operation on transboundary environmental issues in selected priority areas.
>
> **Outcomes**
>
> The "Nile Transboundary Environmental Action" envisaged a number of outputs:
>
> - Enhanced regional co-operation on transboundary environmental and natural resource management issues. Elements include the development and application of a river basin model as part of a decision support system, knowledge management, and linkage of macro and sectoral policies and the environment.
> - Enhanced capacity and support for local-level action on land, forest and water conservation, and establishment of micro-grant fund to support community-level initiatives at pilot sites.
> - Increased environmental awareness of civil society through environmental education programmes and networking of universities and research institutions.
> - Enhanced regional capacity for sustainable management of wetlands and establishment of wetlands management programme at pilot sites.

Case example 5.11. **Transboundary environmental assessment of the Nile basin** (cont.)

- Establishment of standard basin-wide analytical methods for water quality measurements and initiation of monitoring of relevant transboundary hotspots. Enhanced capacity for monitoring efforts and pollution prevention.

Source: www.nilebasin.org, NBI (2001), and edits provided by Asfaw (Project Manager of Nile Transboundary Environmental Action Project) and Hillers (World Bank, AFTSD, Nile Team).

Case example 5.12. **Mekong River Commission Basin Development Plan**

Background and objectives

The Mekong River Commission, on behalf of its member states (Thailand, Viet Nam, Cambodia and Laos), is preparing a Basin Development Plan (BDP) for the Lower Mekong Basin, of which environment is a key cross-cutting theme. The formulation of such a Plan is a key task laid out in the 1995 Mekong Agreement. The Agreement defines the BDP as a planning tool to identify, categorise and prioritise projects and programmes for joint and/ or basin-wide development. It is envisaged that as a planning document, the BDP will contain: a *Basin Development Strategy* and a *Basin Development Management Plan*. The BDP must ensure that harmful effects to other member states resulting from development activities are minimised.

To meet the objectives of the BDP as laid out in the 1995 Agreement, the *Strategy* will need to set out:

- A description of development objectives consistent with the policy of each country.
- An agreed strategy for managing water and water-related resources to best fulfil development objectives.
- A process to identify, categorise and prioritise projects and programmes for joint/basin-wide development.

The *Management Plan* will set out specific actions to develop and manage the basin's resources and the means to monitor these. It has for example been proposed that it will include:

- A portfolio of transboundary programmes and projects to meet strategic needs. These will be made up of:
    - Structural investment projects (*e.g.* bank protection schemes)
    - Non-structural development programmes (*e.g.* regulations to prevent overexploitation of fish stocks).
- Programmes to address identified knowledge gaps (research, etc.).

Approach

SEA should be applied at the strategic level in order to establish the broad environment framework to compliment the BDP Strategy. One of the BDP outputs is a portfolio of basin-wide investment projects. Environmental impacts of the projects should be taken into consideration by using SEA. This could help to determine at which location the different types of projects can best be initiated.

> **Case example 5.12. Mekong River Commission Basin Development Plan** *(cont.)*
>
> At "Level 1 SEA" (during formulation of the Basin Development Strategy), the extent to which broad development interventions (*e.g.* hydropower development, expansion of irrigated agriculture etc.) affect the chosen criteria is assessed. At "Level 2 SEA", long list projects are screened for their potential to cause environmental impacts using a checklist tool. This will determine whether projects will require more detailed project-level Environmental Impact Assessment (EIA) or Cumulative Effects Assessment (CEA) to enable them to move onto a short list or during feasibility stages.
>
> **Outcomes**
>
> The intended outcome from the SEA application will be fed back to refine the Strategy, though some environmental sustainability objectives have already been ensured, including:
>
> - Protection of environment, natural resources, aquatic life and conditions and the ecological balance of the MRC from harmful effects of development (1995 Agreement).
> - Prevention of pollution and other harmful effects of development and acceptance of responsibility for damage caused.
> - Protection of the Tonle Sap lake: development of the river must not impede the natural reverse flow into the Tonle Sap.
> - Maintenance of flows: Mekong flows in both the wet and dry seasons should be maintained within agreed limits (negotiated under the 1995 Agreement).

B. Guidance Notes and Checklists for SEA Undertaken in Relation to Donor Agencies' Own Processes

The following Guidance Notes correspond to the entry points outlined in Table 5.2 above, for which development co-operation agencies play the lead role.

7. Donors' country assistance strategies and plans
8. Donors' partnership agreements with other agencies
9. Donors' sector-specific policies
10. Donor-backed public private infrastructure support facilities and programmes

Guidance Note and Checklist 7: Donors' Country Assistance Strategies and Plans

Description of entry point

In this publication, Country Assistance Strategy/Plan (CAS/P) is used as a generic term for documents that set out the planned programme of assistance provided by a donor to a country, usually for a set period. Such strategies include loan and technical assistance projects, as well as possible co-financing from other development agencies. They are prepared by the agency in close consultation with the government, and often also with other stakeholders, including NGOs.

The terms used to refer to such documents vary across different development co-operation agencies (including "Country Assistance Strategies" at the World Bank, "Country Assistance Plans" at the Asian Development Bank (ADB), "Country Strategic Plans" at USAID, and "Country Strategy Papers" at the EC).

In certain circumstances a CAS/P may be preceded by issue papers outlining a framework strategy for a particularly important element in a country's development processes. For example, a paper on the importance of natural resources to pro poor growth was recently prepared by DFID to help shape its engagement in the Democratic Republic of the Congo, reflecting the importance of natural resources to that country in particular, but lacking a more comprehensive country strategy.

There is a clear role for SEA in developing a CAS/P, preferably aligned with or incorporated into the partner country's domestic procedures. SEAs for individual sector initiatives identified in the CAS/P could then be undertaken by the donor agencies with relevant experience or comparative advantage. But the preferred approach is to support integration of such SEAs as part of national assessment processes. This in turn is dependent on national expertise and competence and might offer opportunities for capacity-building support.

A generic trend to achieve harmonisation of efforts is the development of a Joint Assistance Strategy (JAS). In Tanzania, the government is leading the JAS, aimed at meeting poverty reduction and sustainable development goals, by strengthening national ownership and leadership of the development process, and through consolidation of development partner assistance to the implementation of national strategies, policies and programmes. The JAS is expected to specify all modalities and arrangements of development support to Tanzania and replace individual and multilateral country assistance strategies. This will eliminate multiple development partner processes and requirements and allow for more efficient, effective use of donor resources. Similar processes are beginning in Kenya, Uganda and Zambia.

Rationale: SEA and country assistance strategies/plans

Several assessment approaches have been developed and tested by development co-operation agencies in the context of country assistance strategies and programming, providing useful insights into how an SEA-type approach can be helpful.

One example is the United Nations Development Assistance Framework (UNDAF). This is the common strategic framework for the operational activities of the UN system at the country level. In some countries it also includes Bretton Woods Institutions as well as bilateral agencies. It provides a collective, coherent and integrated UN system response to national priorities and needs within the framework of *a)* the MDGs and other commitments, goals and targets of the Millennium Declaration, and *b)* the declarations and programmes of action adopted at international conferences and summits, and through major UN conventions.

The UNDAF is supported by a Common Country Assessment (CCA) which analyses the national development situation and identifies key development issues. The CCA, which is increasingly an integral part of the UNDAF, includes:

- An assessment and analysis of key development problems and trends, including those addressed by the global conferences and conventions.
- A set of key issues that provide a focus for advocacy and a basis for providing assistance under the UNDAF.

Other examples include the *Environmental Overview* (EO) used by UNDP in the early 1990s, the *Country Environmental Profile* used by the EC to mainstream environment into its Country Strategy Plans, and the *Country Environmental Analysis* used by the Asian Development Bank.

Key questions for donors' Country Assistance Strategies and Plans

Generic questions as well as decisions/activities

- How can sustainable management of natural resources be pro-actively built into proposed programmes in different sectors (*e.g.* health, education, rural development, energy)?
- What are the opportunities for support to environment and natural resource management? What are the respective comparative advantages of different donors?
- What are other development agencies and development banks doing to strengthen environment and natural resource management?
- If budget support is considered as part of the CAS, is there a need for complementary analysis or initiatives to minimize possible negative environmental effects?

> **Key questions for donors' Country Assistance Strategies and Plans** (cont.)
>
> **Linkages/impacts**
> - What currently are the key environmental problems and opportunities and their relation to poverty?
> - What are the linkages between the environment and other important development themes, such as public health (including HIV/AIDS), human rights and democracy, gender, conflicts and vulnerability?
> - What is the importance of environment for pro-poor growth, environmentally sustainable economic development and attaining the MDGs?
> - What are the partner country's commitment to and actual implementation of the Multilateral Environmental Agreements?
> - How are environmental concerns addressed in key partner country strategies, such as the PRS, trade policies and sector strategies, and how are they reflected in the national budget?
>
> **Institutional/implementation**
> - What is the institutional capacity at the national level to integrate environment into planning processes?
> - What donor harmonisation mechanisms are in place to ensure environment is part of donor co-ordination?
> - What are the challenges and opportunities for civil society organisations and the private sector in relation to environment and natural resources management?

Case example

> **Case example 5.13. SEA in Sida's country strategy for Viet Nam**
>
> **Background and objectives**
>
> During 2002-03 a new strategy for Swedish development co-operation with Viet Nam (for the period 2004-08) was produced. The *Vietnamese Comprehensive Poverty Reduction and Growth Strategy* served as a starting point and strategic priorities were identified through analytical work and dialogue with the Vietnamese government and other stakeholders. In line with Sida's policy, an SEA was carried out to ensure the integration of environment into the Country Strategy.
>
> **Approach**
>
> An iterative approach was used to feed environmental aspects into the strategy process at several points:
> - At the initial stages of the strategy process, an environmental policy brief was produced outlining key challenges and opportunities from an environmental and sustainability perspective and linking them to key development issues such as poverty, growth and health.
> - An in-depth environmental and sustainability analysis was produced by a team of WWF-Viet Nam consultants, as one of several background studies conducted as part of the strategy process.

> **Case example 5.13. SEA in Sida's country strategy for Viet Nam** *(cont.)*
>
> - Environment was included as one of several dialogue issues in stakeholder workshops in Viet Nam. The Swedish delegation and Embassy met with government agencies, regional authorities, NGOs, development agencies and other key stakeholders.
> - A workshop was also held with Swedish stakeholders (private sector, civil society, universities and government officials) to discuss the findings from the environmental background study in relation to the country strategy.
> - Detailed comments by environmental specialists were provided on different drafts of the strategy document.
>
> **Outcomes**
>
> - Environment and sustainability issues were well integrated with other important development issues in the final strategy document and following action plans.
> - Key Vietnamese and Swedish stakeholders were involved in the process. As a result of the SEA, stakeholders have deeper understanding of how the environment is intrinsically linked to other critical development issues.
>
> *Source:* Sida.

Guidance Note and Checklist 8: Donors' Partnership Agreements with other Agencies

Description of entry point

A major part of bilateral development agencies' assistance to developing countries is provided through supporting the activities of, and collaboration with, other organisations. This includes formal "strategic partnerships" with multilateral development banks, other multilateral development agencies (such as the European Commission and the United Nations) and other independent organisations with a development or humanitarian mission (such as the International Committee of the Red Cross). The nature, goals and objectives of these relationships are described in Institutional Strategy Papers (ISPs). In addition to formal ISPs, partnership agreements may be developed with Non-Government Organisations and formalised through Partnership Programme Agreements (PPAs) documents.

ISPs and PPAs are strategic and long-term commitments and they provide a framework within which the relationships can be consolidated and deepened for common goals. Their preparation entail consultative processes involving the institutions themselves and a range of civil society contacts. They are frequently supported by logical frameworks, setting out the desired outcomes and outputs and the risks in achieving these. They can then be used to monitor the effectiveness of the relationship.

Rationale: SEA and Donors' partnership agreements with other agencies

Adoption of at least some of the principles of an SEA approach in the formulation and management of these partnerships and their objectives will ensure that environmental considerations are given due attention alongside economic and other considerations. This can help bring consensus and action around the knowledge that sound environmental management contributes to poverty alleviation, while conserving the resource base and environmental services they provide, upon which future livelihoods depend.

SEA principles applied to this process will ensure that agencies focused on development and humanitarian concerns appreciate the important role of environmental and natural resources and, conversely, that those with a focus on environmental issues appreciate the importance of development to environmental protection.

> **Key questions for donors' partnership agreements with other agencies**
>
> **Generic questions as well as decisions/activities**
> - What are the core goals and objectives of the agency and how do they relate to environmental sustainability?
> - Are these compatible with environmental sustainability?
> - Is there adequate appreciation of the relevant environment-poverty links?
> - Does the strategy clearly address environmental threats and opportunities posed directly and indirectly by the agency's activities?
> - How influential is the agency?
> - Does it target sustainable outcomes?
> - How open is the agency to holistic and multi dimensional approaches to decision making?
>
> **Linkages/impacts**
> - Does the agency take account of others working in the area and commit to a co-operative division of labour?
> - Do the agency's networks/organisational structure present opportunities to work with local civil society, government and community based organisations, etc., that the international donor do not have direct access to?
> - Have potential synergies/conflicts/duplication of effort with other agencies been identified and addressed? Is there cross-sectoral coherence?
> - What are the environmental and developmental implications of the key issues addressed by the agency? How will these be managed?
>
> **Institutional/implementation**
> - Are constituent parts of the agency committed to corporate policy and practice?
> - What mechanisms are in place to ensure that environment-poverty links are understood within the agency? Is there a need for capacity building to address sustainability needs?
> - What monitoring and review procedures are in place?
> - Is conditionality/delivering financial support in tranches, etc., an option to encourage change where considered necessary?
> - Is there a management system to address environmental opportunities and risks and to encourage continuous improvement?
> - To what extent does the agency commit to use of local expertise?

Case example

> ### Case example 5.14. **The DFID-WWF Partnership Programme Agreement**
>
> **Background**
>
> WWF-UK – the UK branch of the WWF international environmental organisation – is one of 18 civil society partners supported through the United Kingdom's Department for International Development's (DFID) strategic Partnership Programme Agreements (PPA).
>
> **Approach**
>
> The DFID-WWF PPA provides annually significant, guaranteed and unrestricted funding towards the achievement of mutually agreed long-term outcomes in support of environmentally sustainable development and poverty reduction. The PPA planning and development process is therefore amenable to the application of SEA principles, although this has yet to be done in a rigorous way.
>
> The desired outcomes of the PPA are negotiated between teams in WWF and then with key contacts in DFID. WWF reports to DFID's International Civil Society Division, which provides an opportunity to monitor progress and address any emerging concerns. Formal monitoring requirements for all PPAs have yet to be finalised and it is still unclear as to how these might link to future financial disbursements. DFID and WWF hold annual technically meetings to discuss opportunities for greater "partnership" on a number of key issues (such as climate change) and to better outline how WWF is planning to take forward elements of the PPA. Changing priorities within both agencies require regular liaison.

Guidance Note and Checklist 9: Donors' Sector-Specific Policies

Description of entry point

The importance for development co-operation agencies to apply SEA in the development and implementation of their own sector support policies is increasingly recognised. Such policies set the objectives and measures that will guide or establish a framework for lower tier decisions, such as the preparation of plans and programmes for entire sectors. These are often cross-cutting and integrative in nature, making their environmental assessment complex.

Using SEA in this context can assist in forming a long-term view of the sector and can increase transparency of the sectoral planning process. The SEA can consider the impacts of a variety of planned and unplanned interventions, and determine the additive, synergistic or cumulative effects of discrete activities. Such assessments allow for comprehensive planning of sector-wide mitigation, management and monitoring measures. They also allow the identification of broad institutional resources and technological needs at an early stage. There is now a large suite of tools and increasing experience in applying SEA at the level of sectoral policies.

Rationale: A basis for donor collaboration

SEAs of donor sector policies can provide a basis for collaboration and co-ordination among donors and help avoid duplication of efforts. This is especially important for sector policies that incorporate a range of PPPs.

SEA can help strengthen the subsequent preparation and implementation of sub-projects by recommending criteria for environmental analysis and review, and standards and guidelines for project implementation. It can include analysis of institutional, legal, and regulatory aspects related to the sector, providing the basis for comprehensive and realistic recommendations regarding, for example, environmental standards, guidelines, law enforcement, and training. This way, the need for similar analysis in subsequent environmental assessment work is reduced. SEA helps to alter or eliminate environmentally unsound alternatives at an early stage and identify any areas of conflict/compatibility with existing policies.

Key questions for donors' sector-specific policies

Generic questions as well as decisions/activities

- What are the key drivers for developing or reviewing the sector policy?
- What have been the key linkages with environment and natural resources? Has there been a change in priority (*e.g.* concerns relating to greenhouse gas emissions in the context of energy or transport policies)?
- Does the sector play a major role in meeting MDG targets?
- Has there been an increase in concern for the environmental implications of support to the sector? Or increase in scope for environmental improvement?
- Are conditions in the sector relatively stable and predictable allowing for a medium- to long-term planning horizon?

Linkages/impacts

- Has the policy taken due consideration of all national (donor) policies and other international commitments?
- Has private sector best practice in this sector been considered and incorporated within the sector policies?
- Is there a clear potential for significant environmental improvement or avoidance of major problems in the sector?
- Have synergies with other sectors been considered? Is there cross-sector coherence?
- What are the risks of unexpected outcomes? (Some sector-wide policy changes can affect price signals throughout the economy, leading to a high probability of unexpected outcomes, while others such as education reform are less likely to have unforeseen consequences).

Institutional/implementation

- Have all policy, regulatory, and/or institutional weaknesses affecting environmental management in the sector been identified?
- Do the priority environmental issues imply new institutional mechanisms to address them?
- Is there a need for capacity-building within the donor organisation in order to improve the recognition of environmental linkages?

Case example

> ### Case example 5.15. **CIDA action plan for HIV/AIDS**
>
> **Background and objectives**
>
> The Canadian International Development Agency (CIDA) prepared an Action Plan on HIV/AIDS as part of its social development priorities (2000-05). The plan follows a decision made under the authority of the Minister for International Co-operation and focuses Canadian resources on accelerating progress towards two internationally agreed goals:
>
> - By 2005, at least 90%, and by 2010, at least 95%, of young men and women aged 15 to 24 would have access to the information, education, and services necessary to develop the life skills they need to reduce their vulnerability to HIV infection.
>
> - By 2005, the prevalence of HIV/AIDS in the 15 to 24 age group would be reduced by 25% in the most affected countries, and that by 2010 prevalence in this age group is reduced globally by 25%.
>
> Specifically, CIDA's objectives are as follows:
>
> 1. To work with partners in at least one African country to significantly reduce the number of new HIV cases.
>
> 2. To increase collaboration between CIDA branches and between sectors, sharing lessons and disseminating them more widely.
>
> 3. To increase outreach of CIDA programming to Canadian non-government organisations, academic institutions, and the private sector in order to increase their involvement in international HIV/AIDS work.
>
> 4. To encourage the development of approaches for rapid dissemination in the field.
>
> 5. To increase the quantity and cost-effectiveness of HIV/AIDS interventions funded by CIDA.
>
> **Approach**
>
> *Preliminary issue scan*: The SEA was undertaken in line with CIDA's SEA Handbook. It was first determined that an SEA was required. A statement of the proposal's goals and objectives was produced and a description of the design options prepared. For the purpose of this SEA, the objectives (see above) were taken as the design options and, following a preliminary issue scan, it was decided that only objectives 1) and 5) were likely to have any direct or indirect environmental issues. The rationale for reaching this decision was explicitly described.
>
> *Analysis of environmental effects*: The next step was an analysis of the potential environmental issues of objectives 1) and 5). As the original over-arching HIV/AIDS programme did not yet contain specifics about discrete interventions, it was decided that assessments would be conducted when more specific programmes are defined.
>
> *Stakeholder and public concerns*: CIDA's Handbook recommends that public concerns (both in Canada and the recipient country) should be considered as part of determining the significance of the proposal's environmental issues. Public concerns about HIV/AIDS are already well documented, as are how HIV/AIDS has a detrimental effect on the environment.

> ### Case example 5.15. **CIDA action plan for HIV/AIDS** (cont.)
>
> **Outcomes**
>
> - Documentation, approval and communications: The SEA was documented using CIDA's SEA template.
> - Follow-up and monitoring are part of CIDA's SEA process. Particular environment – HIV/AIDS linkages will be addressed when sub-projects are identified.
> - An overarching SEA for a large programme that currently lacks particular sub-programme activities is beneficial since the SEA identifies programme-environment linkages that must be addressed at the later sub-project level.
>
> *Source:* CIDA.

Guidance Note and Checklist 10: Donor-Backed Public Private Infrastructure Support Facilities and Programmes

Description of entry point

Development agencies provide support to a range of programmes and investment facilities that help build developing country capacity to harness the private sector for better provision of infrastructure services. The aim is generally to involve the private sector to some degree in the financing, ownership, operation, rehabilitation, maintenance or management of infrastructure services. Most of these facilities and programmes are multi-donor supported but managed and operated by a single donor or an organisation set up for the purpose.

Development agencies, as shareholders or managers of these funds or facilities, need to be able to ensure that the specific activities and investments funded are subject to effective environmental risk management. In part, this is driven by the need to ensure compliance with their own environmental policies (even though the investments are being managed by another party). The broader aim is to build sound sustainability objectives into local infrastructure investment for the benefit of the community.

The SEA approach that can be undertaken is twofold. The first element entails the strategic environmental audit of the "footprint" of an existing investment facility, in order to assess the current exposure of the development agency to any risks arising from the environmental effects of the investment portfolio. The second element is to strengthen the capacity of the manager of the facility to mainstream environmental risk assessment into the process for appraising, approving and monitoring investments.

Rationale: donor-backed public private infrastructure support facilities and programmes

An SEA can be applied to the portfolio of existing investments, analysing their environmental and social footprint. Such an audit would require the analysis of all investments, looking at the cumulative environmental issues and corresponding positive opportunities and negative aspects. This is an approach suited to the requirements of a development agency that is a shareholder in a facility but not directly responsible for managing it. It addresses past performance and the resulting reputation risk to the development agency. Furthermore, it allows for the identification of indicators for monitoring environmental outcomes, which in turn can influence the criteria for appraising future investments.

An environmental risk management system can be designed as part of the management of the infrastructure programme to meet the environmental policy requirements of the shareholders. The system would have the strategic objective of ensuring that the governance processes, investment appraisal procedures and management controls developed to run the facility fully integrate the aim of ensuring environmental sustainability. This is to ensure effective environmental and social risk management of investments. Such an approach requires effective governance, but results in a solid framework of accepted principles of social and environmental responsibility. This approach has been successfully developed by the IFC and EBRD over the past decade or so to ensure effective environmental risk management at the level of Financial Intermediaries through which much of the financial support is channelled. It has also been deployed in the context of community-driven development – long-term programmes financed by development agencies to support small scale locally managed infrastructure investments.

Key questions for donor backed public private infrastructure support facilities and programmes

Generic questions as well as decisions/activities

- What is the policy environment within which these programmes will take place?
- What are the objectives of the facility or programme? Are MDGs, poverty reduction, and environmental sustainability targeted?
- What is the profile of the investments expected? Will they occur in sectors with clear environmental risks and opportunities?
- How sensitive will the success of such investments be to environmental sustainability and community support?
- Are there clear environment-related opportunities to be gained by channelling investment into the infrastructure sector?

Linkages/impacts

- What are the key environmental and social issues typically associated with the relevant sector (*e.g.* water services, energy, and transport)? Are there climate change-related concerns?
- Are there well-established principles of sound environmental management that can be built into the strategy for the facility?
- Is there an effective environmental risk management system that can be applied at the individual project level, with screening criteria, guidance on assessment and suggested environmental mitigation and management measures?
- Has the facility drawn lessons from the experience of other private sector investments?

Institutional/implementation

- Do the governance arrangements for the investment facility make clear the environmental and social sustainability goals and processes?
- Do the procedures and management controls for the facility incorporate measures to ensure the environmental risk management system is applied and resourced effectively?

> **Key questions for donor backed public private infrastructure support facilities and programmes** *(cont.)*
>
> - Does the staff have sufficient knowledge and experience to put it into practice?
> - Are appropriate monitoring indicators and processes in place? Are post-implementation reviews and benchmarking planned?

Case examples

> **Case example 5.16. Environment due diligence for financial intermediaries, based on European Bank for Reconstruction and Development (EBRD) procedures**
>
> **Background**
>
> The development agency that provides finance through a Financial Intermediary (FI) (for example a local bank) faces the challenge of ensuring that the FI develops and applies a sound environmental risk management system to guarantee that the on-lending to individual companies meets the development agencies overall environmental management objectives. Environmental Due Diligence' should be an integral part of the FI's credit appraisal process, because environmental risk may translate into credit risk. If a client company of the FI has environmental problems, this may affect its ability to honour loan repayments or contractual obligations, for example because:
>
> - The company has to make major investments in order to meet regulatory requirements.
> - The company's activities are scaled down, suspended or closed down by the local authorities due to environmental problems.
>
> **Objectives**
>
> The Environmental Due Diligence process will help clients of the Financial Intermediary to:
>
> - Ensure compliance with EBRD's environmental requirements.
> - Identify any environmental issues associated with a particular client/transaction.
> - Identify and evaluate the financial implications related to environmental issues.
> - Minimise exposure to environmental/financial risks.
> - Maximise opportunities for environmental benefits and minimise the potential for adverse environmental impacts (such as pollution) associated with clients.
> - Protect the client and development agency from reputation risk associated with financing companies with a poor environmental record.
>
> **Tools**
>
> EBRD due diligence questionnaire for FI clients operating in environmentally sensitive sectors:
>
> 1. Nature of the client's business (name, location, industry sector, product manufactured, capacity, number of employees, main markets).
> 2. Environmental regulatory compliance and liability:
>   - Does the company comply with national health, safety and environmental (HSE) regulations and standards?
>   - Is the company in possession of all required HSE permits and approvals?

> **Case example 5.16. Environment due diligence for financial intermediaries (Based on European Bank for Reconstruction and Development (EBRD) procedures** (cont.)
>
> - Do the products manufactured by the company meet national HSE regulations and product standards as well as those of the countries of export?
> - Has the company paid excess charges or fines/penalties for non-compliance with HSE regulations and standards in the last two years?
> - Is the company subject to on going or pending administrative or court action because of environmental offences?
> - Is the company exposed to potentially significant HSE liabilities, such as those arising from land or groundwater contamination, related to the company's past or on-going operations? If yes, specify magnitude.
>
> 3. Has the company had any significant accidents or incidents in the last two years (*e.g.* oil spills, fires) involving deaths or multiple serious injuries and/or significant environmental damage?
>
> 4. In the event that the company is not materially in compliance with HSE regulations and standards, or if there are potentially significant HSE liabilities, please describe further actions required by the authorities and/or planned by the Company to address these issues satisfactorily, and to achieve regulatory compliance.

> **Case example 5.17. Environmental risk management of a community-driven development project – Programme National de Développement Participatif (PNDP) Cameroon**
>
> **Background**
>
> Development agencies increasingly provide support to poor rural communities through decentralised, locally managed funds for small-scale infrastructure and natural resource management projects – Community Driven Development. The development agency faces the challenge of ensuring that the fund is administered in a manner that addresses environmental risks and opportunities that complies with its own environmental safeguard policies. The potential solution is to build an Environmental and Social Management Framework (ESMF) into the design of the fund.
>
> **Objectives**
>
> The purpose of the Programme National de Développement Participatif (PNDP) is to reduce poverty and promote sustainable development in rural areas of Cameroon. The PNDP aims to support community-driven development by allowing communities and their local government ("communes") to implement priority action plans. This is achieved through strengthening the fiscal, institutional, and administrative environment for adequate budget allocation, effective service delivery, and transparent management of financial services. A key component will co-finance (through grants) collective socio-economically beneficial micro-projects such as social infrastructure and natural resource management activities.

> **Case example 5.17. Environmental risk management of a community-driven development project – Programme National de Développement Participatif (PNDP) Cameroon** *(cont.)*
>
> **Tool**
>
> An Environmental and Social Management Framework (ESMF) was developed which includes:
>
> - A screening tool for micro-projects.
> - An exclusion list.
> - Guidance on compliance with safeguard policies at micro-level.
> - Guidance on identifying environmental risks and opportunities.
> - Implementation and reporting requirements.
>
> **Procedure**
>
> - At the provincial level, a full-time Environmental and Social Mitigation Officer (ESMO) in each Provincial Project Unit (PPU) will be appointed to provide technical backstopping on all aspects of environmental and social mitigation in line with the ESMF.
> - These ESMOs will also be trained in the project's Resettlement Policy Framework in order to support the communes in identifying and promoting sustainable practices for land management, land tenure, land acquisition and involuntary resettlement and conflict resolution.
> - An annual environmental and social performance audit will be carried out by an independent consultant.
> - The ESMOs will work with the Commune Decisions Committees (CDCs) and an ESMF facilitator to develop strategic approaches to environmental sustainability in their communities.
> - At more practical levels, specific studies would be carried out on issues of environmental and social management assessment, and the details of the ESMF will be integrated into the micro-project cycle.

C. Guidance Notes and Checklists for SEA in Other, Related Circumstances

The following Guidance Notes correspond to the entry points outlined in Table 5.4 above, for which donors, while involved, do not play a lead role.

11. Independent Review Commissions (which have implications for donors' policies and engagement)
12. Major private sector-led projects and plans

Guidance Note and Checklist 11:
Independent Review Commissions
(which have implications for donors' policies and engagement)

Description of entry point

There have been a number of independent, international, sector-based Review Commissions such as the World Commission on Dams and the Extractive Industries Review. By their nature, these undertake a process analogous to SEA. They carry out comprehensive reviews of their sectors using multiple levels of analysis and review. Such Commissions review the global performance and impacts of the sector, providing an integrated analysis. They involve significant multi-stakeholder engagement and can help to develop recommendations for the donor institutions' future involvement in the sector.

The approaches employed are comprehensive, integrating social, environmental and economic dimensions of development. As with conventional forms of SEA, potential strategic alternatives are assessed (both the opportunities they provide, and the obstacles they face), as well as governance and institutional processes. Planning, decision-making and compliance issues that underpin project selection are also analysed. Such reviews will identify examples of best practice and provide recommendations for future planning and decision making.

Rationale: Common elements in the approach of Independent Review Commissions and SEA

The process undertaken by Review Commissions is analogous to conventional SEA. The following list of steps is not exhaustive, and different review bodies will employ different levels of analysis, but the commonalities between the two approaches are clear. In general, a review will:

- Identify and assess the environmental consequences of the sector using multiple levels of analysis and review.
- Assess potential alternatives to the developments.
- Review the governance and institutional processes.
- Facilitate and respond to consultation with stakeholders.
- Improve the evidence base for strategic decisions.

- Contribute to more transparent planning, involving all stakeholders and integrating environmental considerations.
- Contribute to the goals of sustainable development.
- Identify examples of best practice.
- Provide relevant predictions about the future of the sector.
- Develop recommendations for future involvement in the sector.
- Monitor environmental effects resulting from activities.

Key questions for Independent Review Commissions

Generic Questions as well as decisions/activities

- How compatible is engagement in the issue/sector under consideration with the goals and policies of the agency under review?
- What are the feasible strategic choices open to consideration?
- How influential is the agency in setting benchmark performance standards?
- Is there evidence that engagement/investment in this issue or activity further development goals and achieve sustainable outcomes?
- Is continued engagement/investment in this issue or activity to be encouraged, or is this counter productive to development goals?
- What are the environmental and social risks and opportunities of on-going engagement in the sector/issue?
- Is there a net environmental benefit from continued engagement?
- If continued engagement has unrealised potential benefits, are there conditions or principles of engagement that need to be put in place to realise this potential?
- Does decision making reflect a balance between all the dimensions of sustainability?

Linkages/impacts

- What are the typical key issues of concern, and how has the relevant agency responded to these in the past?
- What are the implications of continued investment/engagement for international commitments to multilateral environmental agreements, national legislation, policies and commitments in host countries, etc.?
- Does practical implementation at the project level demonstrate the agency's(ies) policy intentions?
- What is an adequate representation of geographical-, scale-, etc., differences from which generic observations can be developed?
- What have been the distributional impacts of the investments in the issue/sector in the past?

Institutional/implementation

- Is the Review body itself reflective of all the key stakeholders?
- Is the Review process accessible and transparent?
- Is impartiality of the Review body demonstrated?
- Are all relevant stakeholders ordinarily able to access key decision makers in the agency(ies)? If not, should they be given preferential hearing during the Review process?

II.5. APPLICATIONS OF STRATEGIC ENVIRONMENTAL ASSESSMENT IN DEVELOPMENT CO-OPERATION

> **Key questions for Independent Review Commissions** (cont.)
> - Does the Review body comprise sufficient authoritative expertise, so that the agency(ies) recommendations/ findings will be seriously considered?
> - Are the elements of good governance in place in the agency(ies) under review?
> - Is there commitment by the agency(ies) top management to implement findings of the Review body?
> - Is the Review body independent of the agency(ies)?
> - Is a monitoring process in place to track implementation of the accepted recommendations of the Review?

Case examples

Case example 5.18. The World Commission on Dams

Background and objectives

The World Commission on Dams (WCD) – an independent, multi-stakeholder process sponsored by the World Bank and IUCN – was established to review the development effectiveness of large dams. Its brief included the consideration of alternatives for water and energy services; developing internationally acceptable criteria and guidelines for planning, designing, construction, operation, monitoring, and decommissioning of dams. The WCD reported its findings in November 2000.

Over two years, the WCD conducted the most comprehensive, global and independent review of large dams to date. The World Bank said it would use the WCD report "as a valuable reference … when considering projects that involve dams".*

Approach

The WCD mandate was to:
- Review the development impact of large dams and assess alternatives for water resources and energy development.
- Develop international criteria, guidelines and standards, where appropriate, for the planning, design, appraisal, construction, operation, monitoring and decommissioning of dams.

The WCD undertook the following to develop a knowledge base upon which to base its recommendations:
- In-depth case studies of large dams in five continents, together with two country papers.
- A cross-check survey targeted at 150 large dams in 56 countries.
- 17 thematic reviews.
- Four regional consultations.
- Inputs submitted by interested individuals, groups and institutions.

Outcomes

The global review had three components:
- An independent review of the performance and impacts of large dams (looking at technical, financial and economic performance; ecosystem and climate impacts; social impacts; and the distribution of project gains and losses).

II.5. APPLICATIONS OF STRATEGIC ENVIRONMENTAL ASSESSMENT IN DEVELOPMENT CO-OPERATION

> ### Case example 5.18. **The World Commission on Dams** (cont.)
>
> - An assessment of the alternatives to dams, the opportunities they provide, and the obstacles they face.
> - An analysis of planning, decision-making and compliance issues that underpin the selection, design, construction, operation and decommissioning of dams.
>
> The Commission developed criteria and 26 guidelines. These were developed to help States, developers and owners, as well as affected communities and civil society in general, meet emerging societal expectations when faced with the complex issues associated with dam projects. These aim to foster informed and appropriate decisions, thereby raising the level of public acceptance and improving development outcomes.
>
> * World Bank's policy is detailed and explicit for international waterways. It is less so in other areas of potential transboundary effects. World Bank and the World Commission on Dams Report, Q&A, March 2001.
>
> *Source:* World Commission on Dams (2000).

> ### Case example 5.19. **The Extractive Industries Review**
>
> **Background**
>
> In response to concerns that the World Bank's ongoing investment in the extractive industry sector (oil, gas and mining) was incompatible with its broader commitments to poverty reduction and sustainable development, the World Bank commissioned an independent Extractives Industries Review (EIR).
>
> The basic premise underwriting the EIR was that the extractive industries sector has the potential to be the engine of economic growth in many developing countries. Paradoxically, however, developing countries that are reliant on exploiting significant mineral resource endowments often exhibit greater degrees of corruption, poverty, conflict and poor governance (along with more project specific negative environmental and social impacts) than those that do not have such mineral resources.
>
> The 2004 EIR report *Striking a Better Balance – The World Bank Group and Extractive Industries* concluded that the WBG should stay engaged with the sector, but only where its investments can be seen to explicitly support poverty reduction and sustainable development goals, and when three critical enabling conditions are in place: pro-poor public and corporate governance, more effective social and environmental safeguard policies, and greater respect for human rights.
>
> **Approach**
>
> In order to consider the critical strategic choice of whether the WB should stay engaged in the extractives sector, the EIR carried out an extensive review of existing project experiences, invited written submissions, conducted interviews, and held five regional workshops to hear evidence.
>
> The review process attempted to accommodate multi-stakeholder interests, but was criticised for the inordinate influence it gave to civil society in the consultation process. This was countered by the EIR's contention that civil society does not normally have direct, continuous and official links to operational WBG decision making on a par with the participation of governments and the private sector. Thus a need was identified to design a civil society bias into the process.

> **Case example 5.19. The Extractive Industries Review** *(cont.)*
>
> *"The WBG needs to know that genuine development requires partnership not only with governments and companies, but with civil society as well ... It is civil society – local communities, indigenous people, women and the poor – who suffer the negative impacts of extractive industrial development – such as pollution, environmental degradation, resettlement and social dislocation."*
>
> The EIR process was transparent. All submissions, draft report iterations, representations, developments, etc., were posted on the EIR Web site (*www.eireview.org*). In addition, the recommendation highlighted the principle of transparency as a key dimension of good governance, with cross-reference to the Aarhus Convention's three pillars of access to information, public participation in decision making and access to justice (UNECE 2000). The EIR also explicitly addressed the interrelationships between social, economic and environmental aspects.
>
> **Outcomes**
>
> The report recommended that *"a strengthening of the environmental and social components of WBG interventions"* was required.
>
> It added that the WBG:
>
> *"should take an holistic, multi-dimensional approach to assessments, identifying cumulative impacts of projects and socio-economic linkages to environmental issues"*
>
> and that the WBG should recognise the need to:
>
> *"mainstream economic, social and environmental considerations into sustainable development – with the alleviation of poverty as the economic goal, the enhancement of human rights as the social goal and the conservation of the ecological life support system as the environmental goal"*.
>
> **Further, the EIR recommended the use of strategic assessments:**
>
> *"The structural framework within which the oil, gas and mining sectors exist is of fundamental importance to achieving pro-poor development outcomes that are sustainable. Poverty and the environment should be accorded strategic importance in designing and implementing structural reform programmes that include the extractives industries."*
>
> Implementation of the World Bank's managements response to the EIR is the subject of regular progress reports to the WB Board and overseen by an Advisory Committee of external stakeholders.
>
> *Source:* EIR (2004).

Guidance Note and Checklist 12:
Major Private Sector-Led Projects and Plans

Description of entry point

The private sector has a huge impact on development outcomes – through its investments and its interactions with other partners in shaping and stimulating development. Wherever the private sector considers investment options or commercial strategies in a developing country, it has the potential to affect development prospects significantly – positively and negatively. Also, in almost every situation, there are potential overlaps in the interests and activities of the private sector and the development agencies working to reduce poverty and encourage growth.

Both the private sector and development agencies need to make strategic choices – about how investments are channelled and what types of investment can deliver the best developmental benefits. Equally, in making strategic choices, actors can weigh up the environmental and social implications of these choices. Scenario development and analysis is a tool that can specifically help in this task (see Annex C, Section 2.2).

In many cases, the harmful social and environmental effects of private sector – led strategies and investments, suggest that an SEA undertaken prior to the investment would have enabled investors to consider better options at an earlier stage. Several recent high-profile oil, gas, mining and water resource investments illustrate this point.

Development agencies (such as EBRD or IFC) are often involved alongside the private sector, financing the host government's contribution to a project or as shareholders in financing institutions They and their private sector partners have a joint interest in applying SEA so that the highly significant environmental issues are considered as part of the early planning process before specific project details are considered.

Accordingly, there are increasingly more examples of the private sector applying SEA-type approaches, usually for large multiple project investments with high social and environmental risks, not limited to the narrow scope of a specific project. Undertaking such an assessment provides them with greater certainty about the scope and limits of future development, identifies future risks and opportunities, and can demonstrate that decision making has been transparent and balanced. The participation of affected groups during, rather than after, the formative decision-making process is also essential for responsible operators.

Rationale: SEA applied to major projects led by the private sector: The example of major oil and gas extraction and transportation programmes

The Oil and Gas sector provides a good example of where SEA benefits private sector major projects. Developments in this sector typically consist of a set of integrally related projects that together constitute the extraction, refining and transportation infrastructure.

In recent years there have been many controversies regarding the impacts of major oil and gas extraction and transportation programmes. In particular, there has been criticism that the social and environmental impacts of these developments have been inadequately considered, and that local groups and communities have not been adequately engaged and consulted before decisions to proceed were made. SEA can be used to address these problems.

Key questions for major private sector-led projects and plans

Generic questions as well as decisions/activities

- Has the country's policy and programme for the relevant sector been reviewed and, where appropriate, taken into account in the design of the investment?
- Have all feasible options and alternative plans been identified and compared to the proposed investment?
- Has international experience of issues and challenges been reviewed?
- Have all relevant stakeholders at national and international level been consulted about the investment strategy?

> **Key questions for major private sector-led projects and plans** (cont.)
>
> **Linkages/impacts**
>
> - What are the key linkages of the investment strategy to the environment and social issues? What are the key risks and opportunities?
> - What are the implications for global environmental, issues such as greenhouse gas emissions or biodiversity?
> - What are the related social and economic effects likely to result from the investment strategy?
> - What are the broader and cumulative effects that may affect the context for the strategy?
> - Have potential transboundary effects been identified? If yes, has notification/information exchange taken place prior to major strategic decisions for the investment?
>
> **Institutional/implementation**
>
> - Are there any market, policy or institutional failures that need to be considered in project design?
> - Is there government commitment to sustainable private sector involvement? Are the fundamentals of good governance in place to encourage business confidence?
> - What is the level of public concern at the local, national and international levels about the project? Are public engagement activities in line with government policies? Is it part of the policy process?
> - Who are the key implementing bodies for the project? Is there need for capacity building for those responsible for implementing the investments?
> - Are appropriate monitoring indicators and processes in place? Are there post-implementation reviews and benchmarking planned?

Case examples

> **Case example 5.20. Nam Theun 2 Hydropower Project, Lao PDR**
>
> **Background and objectives**
>
> The Nam Theun 2 project is located in the Khammouane and Bolikhamsay provinces in Central Lao Peoples' Democratic Republic (PDR). The project includes the development, construction, and operation of a thousand-megawatt hydropower facility primarily for export of power to Thailand. It includes a 450 km^2 reservoir on the Nakai Plateau and will divert water from the Nam Theun basin to the Xe Bangfai basin. The project is privately financed, but with involvement from the World Bank and ADB who have engaged with the investors to agree the level of strategic and cumulative impact assessment to be undertaken.
>
> The hydropower sector seems to be one of the most thoroughly planned sectors in Lao PDR from an economic and technical viewpoint. However, there is also strong awareness that hydropower developments can induce:
>
> - Direct or indirect impacts caused by the changes in river morphology, hydrology and ecology (that can occur a long way upstream and downstream).

Case example 5.20. **Nam Theun 2 Hydropower Project, Lao PDR** (cont.)

- Direct and indirect changes caused by the necessary large-scale construction work.

and that such developments often take place in remote and pristine areas where local people will bear the negative impacts but might see little benefit from increased power production.

Approach

The proposed Nam Theun 2 Hydropower Project would have sector-wide implications including environmental and social impacts. A strategic approach to management of environmental and social impacts in the sector was needed to consolidate, update and expand previous work related to hydropower and environment, and to clarify the broader issues faced by hydropower development in Lao PDR. The SEA had a broader focus than the extensive project level Environmental and Social Impact Assessment undertaken by the developers. It looked at sector-wide issues and cumulative and transboundary impacts. It recommended strategic actions to improve environmental and social management in support of the implementation of the Hydropower Development Strategy including the Nam Theun 2 Hydropower Project.

Outcomes

The SIA covered planned hydropower developments in Lao PDR from a 20-year perspective. It will contribute to a better understanding of the impacts of hydropower development plans in Lao PDR and recommend measures strengthening the sector in order to reduce impacts and manage the sector in an environmentally and socially sustainable manner.

Source: World Bank and Norplan (2004).

Case example 5.21. **The potential of SEA in relation to major oil and gas investments**

Sakhalin Island, Russia: There are a number of blocks around Sakhalin Island held by a variety of operators. The cumulative impacts of these oil and gas infrastructure projects have not been assessed together, resulting in multiple pipeline systems and no clear picture of the combined impact of the component parts of the development. The progress of the individual investments is being harmed by the absence of a strategic assessment.

Baku-Tbilisi-Ceyhan (BTC) pipeline: No formal SEA has ever been undertaken for the proposals for the BTC pipeline or the wider Caspian Sea oil and gas developments, meaning that a whole range of alternatives for, and the cumulative impacts of, the different components of the project have never been assessed. Even the Environmental and Social Impact Assessment conducted for BTC was split into three separate sections, limiting the options for consideration.

Norwegian management of the Barents Sea: The Norwegian government recognised the importance of protecting the Barents Sea ecosystem and other marine areas and is developing integrated management plans for its coastal and marine areas, starting in 2002 with the Barents Sea. The plan will address the impacts of fishing, aquaculture, oil operations and shipping. It will attempt to ensure that the accumulated effect on the ecosystem does not exceed the tolerance of the ecosystem, and that the strategic, integrated approach inherent to SEAs is adopted.

> **Case example 5.21. The potential of SEA in relation to major oil and gas investments** *(cont.)*
>
> **UK Offshore Oil and Gas Industry:** The Department of Trade and Industry (DTI) is the principal regulator of the offshore oil and gas industry in the UK. It used SEA proactively to strike a balance between promoting economic development of the UK's offshore oil and gas resources and effective environmental protection. In 1999, the DTI began a sequence of sectoral SEAs of the implications of further licensing of the UK Continental Shelf (UKCS) for oil and gas exploration and production (before the EU SEA Directive came into effect in 2004). The UKCS has been subdivided into eight areas. To date five Licensing Rounds have been subject to separate SEA (see also *www.offshore-sea.org.uk*).

Notes

1. A companion resource book (Dalal-Clayton and Sadler 2005) is available at *www.seataskteam.net*. This provides case studies in four categories: developed countries, development co-operation, countries in transition and developing countries.
2. Under World Bank Operational Policy 4.01.
3. The World Bank undertakes an unofficial review of PRSP documents as they are submitted, using a scoring system. Based on this approach, the Southern Africa Institute for Environmental Assessment has developed a quantitative analysis of poverty/environment linkages and integration in PRSPs. (Available at *www.seataskteam.net*.)
4. UNDP/UNEP/IIED/IUCN/WRI, 2005.
5. Adapted from DFID/EC/UNDP/World Bank (2002).
6. The World Bank instrument replaced macro-economic structural adjustment lending in 2004.
7. The publication *Environmental Fiscal Reform for Poverty Reduction*, 2005, DAC Guidelines and Reference Series, provides more detail on these issues.

Part III

Chapter 6. **How to Evaluate Strategic Environmental Assessment** 123

Chapter 7. **Capacity Development for Strategic Environmental Assessment** . . . 129

Part III

Chapter 6 How to Evaluate Strategic Environmental Assessment   123

Chapter 7 Capacity Development for Strategic Environmental Assessment   209

ISBN 92-64-02657-6
Applying Strategic Environmental Assessment
Good Practice Guidance for Development Co-operation
© OECD 2006

PART III

Chapter 6

How to Evaluate Strategic Environmental Assessment

Officials in development co-operation agencies are more likely to have to review an SEA process and its results than actually undertake it themselves. This chapter sets out key questions that will help them in this role. Evaluation is important to determine whether the outcomes have been achieved, fully or in part, and also to ensure quality control of the SEA process itself. An evaluation of an SEA can be limited to the relatively easy task of verifying whether the SEA proposed more sustainable alternatives. It is more ambitious to determine whether the SEA led to a more sustainable PPP design and implementation. This requires extending the focus to include the effects on institutional and capacity-building issues which highly influence the implementation process.

6.1. The role of evaluation

Evaluation examines whether an intervention has achieved its intended outputs and outcomes. The challenge is to define clearly how to measure these achievements in an objective and robust manner. This approach needs not be too complicated – there may be elements that can be measured more objectively than others, especially where cause-effect relations are difficult to determine with any level of certainty. Evaluation of an SEA is likely to involve examination of cause-effect "plausibilities" to some degree – an informed judgement about whether an SEA did or did not finally influence the design, planning or decision about a PPP.

A systematic approach to evaluation (and monitoring) can be supported by a list of questions as set out in this chapter. The important point of evaluation is not to seek absolute scientific proof but to engage in reflective processes to evaluate and improve on previous decisions. In this way, the aim is to learn how to continuously improve the integration of sustainability dimensions into decision making, and how to improve the use and efficiency of an SEA as an approach for sustainable development.

In this context, evaluation of an SEA can also help to:

- Improve learning on the linkages between PPP formulation/assessment and their practical outcomes.
- Achieve PPP goals by identifying *ex post* adaptation requirements for those implementation mechanisms/actions that have failed to deliver their intended outcomes.
- Support the accountability of decision makers and involved stakeholders by making the results of decisions transparent.

A central element of evaluation is the definition of appropriate indicators that reflect sustainable outcomes as a result of implementing the PPP. Indicators are also essential to quantify the achievement of specific objectives and goals. Appropriate indicators should be defined during the SEA process to enable the necessary data to be collected during the implementation phase.

Some aspects of objective and goal achievement are better evaluated in a qualitative manner. Hence, written descriptions of the envisaged objectives can be compared with what was practically achieved. Checklist 6.1 provides a preliminary list of such questions.

Evaluation should not be an academic exercise. Ideally, it should lead to concrete results which might include:

- Positive recommendations on future actions.
- *Ex post* adaptation of implementation measures, or even of the PPP decisions – these will be inevitable if serious deviations from previous assumptions endanger the achievement of specific goals.
- Specific measures to develop capacity, tailored to help overcome implementation gaps.

Subsequent donor support might be designed to reflect the outcome of an evaluation.

6.2. Evaluating the delivery of envisaged outcomes

Perhaps the most important outcome of a good quality SEA is that it has significantly influenced the achievement of positive development results and has helped to enhance the effectiveness of development. But development involves complex processes and it is not easy to isolate those outcomes that are solely due to the application of SEA (attribution gap). Equally, it is impossible to ascertain whether unsustainable outcomes of a PPP would have been avoided by undertaking an SEA.

Checklist 6.1 provides questions to help evaluators focus on important outcome aspects of an SEA.

Checklist 6.1. Key questions for evaluating the delivery of envisaged outcomes of a PPP

Assumptions made during the SEA

- Did the SEA predict future outcomes correctly? Were the assumptions made during the SEA for modelling expected impacts and/or institutional and governance requirements correct?

Influence of the SEA on the PPP process

- Did the SEA provide useful information for those responsible for developing the PPP?
- Did the SEA identify the issues most important to sustainable outcomes, rather than all significant environmental issues?
- Did the SEA reflect questions and concerns not initially included in the PPP? What was appreciated most/what was irrelevant, etc.?
- Could the SEA findings be effectively conveyed to the decision makers?
- Were the decision makers willing to reflect on and include the provided information in decision making?
- Did the SEA succeed in actually changing the PPP/making the PPP more environmentally sound?
- Did the PPP process make sufficient reference to the findings of the SEA?

Influence on the implementation process

- Did the SEA succeed in actually changing the PPP implementation or budget plans, or other subsequent measures, making the PPP more environmentally sound?

> **Checklist 6.1. Key questions for evaluating the delivery of envisaged outcomes of a PPP** *(cont.)*
>
> - Did the PPP actually lead to implementation measures and outcomes that better reflect the goals of sustainable development/environment? Were options implemented which were more environmentally sound?
> - Did the recommendations of the SEA lead to change in institutional settings (*e.g.* an advisory group on environment, inter-sectoral co-ordination, subsequent EIA requirements, etc.) and governance (for example access to judiciary or empowerment of weak stakeholders for environmental management) which supported the integration of sustainable development/environment during implementation?
> - Did different stakeholders of relevance for the implementation act on recommendations by the SEA during the implementation process?
>
> **Influence on direct and indirect goals of relevance to sustainable development/ environment**
>
> - Are there any indications that the SEA contributed to:
>   - The achievement of MDG 7 and/or other goals of relevance in the particular case?
>   - Improved conditions of environment and natural resources in the relevant area?
>   - Transparency and accountability, and improved governance?
> - Did the sustainable development benefits of the SEA outweigh the costs associated with carrying it out?
>
> **Outcome on capacity building and influence on accountability**
>
> - Did the SEA help build capacity by training decision makers or implementers?
> - Did the SEA empower weak and vulnerable stakeholders?
> - Did the SEA enhance the transparency of decision-making processes and accountability of decision makers on the environmental implications of PPPs?
> - Did decision makers justify or correct their decisions based on SEA findings and monitoring?
> - Did the application of SEA lead to a better understanding of the potential of this approach and, possibly, encourage SEA applications later on?

6.3. Evaluation as quality control check

In a formal sense, a "good SEA" is one that conforms to the key principles listed in Chapter 4. These are elaborated in Checklist 6.2 to help those engaged in reviewing an SEA process to gauge success. This task should be carried out throughout the SEA process. Taken cumulatively, the lessons from such process evaluation will influence the evolution of SEA practice in development co-operation.

> **Checklist 6.2. Key questions for evaluation as a quality control check**
>
> **Presentation and quality of information**
>
> - Was the information provided by the SEA process adequate (*i.e.* comprehensive, rigorous and understandable) from the point of view of those responsible for developing the PPP? What was missing?

Checklist 6.2. **Key questions for evaluation as a quality control check** (cont.)

- Was the information provided by the SEA process adequate (see above) from the point of view of the key stakeholders? What was missing?

Co-operation and stakeholder participation

- Has there been effective co-operation between the SEA team and those responsible for developing the PPP? Why? How can this be improved?
- Was there effective public involvement? Why? How can this be improved?
- Was there an effort to involve less powerful stakeholders in the consultation? If so, how successful was this?

Description of the SEA procedure in the report

- Has the purpose/aim of the SEA been described with a mention of the regulations which underpin the SEA process and document?
- Is the scope of the SEA discussed?

Objectives used for the SEA

- Have the substantial objectives used for the SEA been described and defined, quantitatively where appropriate?
- Does the SEA report identify and describe any conflicts that exist between the objectives and the PPP, and between the objectives and other PPPs?

Alternatives

- Are the potential alternatives within the PPP described and considered in terms of the SEA objectives? Have these included the "no change" alternative?
- If any alternatives have been eliminated, have the reasons been provided?

Assessment of environmental impacts

- Where there are likely to be significant environmental affects, are they clearly described?
- Is an effort made to prioritise those effects that most affect sustainability?
- Are the methodologies for assessing environmental impacts described?
- Is the full range of positive and negative impacts addressed?
- Where there are uncertainties in assessing the impacts and assumptions have been made, have they been justified and the worst-case scenario used?
- Are mitigation measures clearly described and committed to that will prevent, reduce or remedy any significant adverse effects on the environment in implementing the PPP?

Planned follow up activities and implementation

- Are the indicators for monitoring clearly defined? And, are they based upon the original baseline information and on the objectives of the PPP and the SEA?
- Are the links to other potential follow-up procedures specified, *e.g.* project EIA, design guidance, etc.?
- Are recommendations for the implementation process clearly formulated?
- Are outcome indicators defined? And is there an evaluation plan (with adequate budget and clearly assigned responsibilities) so that the sustainability focus of the SEA can continue beyond the planning phase?

> **Checklist 6.2. Key questions for evaluation as a quality control check** *(cont.)*
>
> **Overall comments on the SEA process**
> - What is the view of key stakeholders (particularly the less powerful ones) and those responsible for developing the PPP on the different elements of the SEA?
> - How could it be improved in future?
>
> **Constraints and opportunities**
> - What were the most significant constraints to achieving an effective SEA?
> - What were the most significant positive factors ensuring success of SEA?
>
> *Based on evaluation criteria prepared by Rasso (2002) and the Institute of Environmental Management and Assessment (www.iema.net).*

ISBN 92-64-02657-6
Applying Strategic Environmental Assessment
Good Practice Guidance for Development Co-operation
© OECD 2006

PART III

Chapter 7

Capacity Development for Strategic Environmental Assessment

III.7. CAPACITY DEVELOPMENT FOR STRATEGIC ENVIRONMENTAL ASSESSMENT

This chapter discusses the importance and role of developing capacity in the application of SEA both within development co-operation agencies and in partner countries. It focuses on capacity in terms of skills and institutional needs, in relation to both the development of SEA systems and SEA application and evaluation. Examples of SEA capacity-building initiatives are included.

7.1. Why is SEA capacity development needed?

Practical experience of applying SEA has highlighted two key challenges:

- Lack of knowledge amongst decision makers and relevant administrations regarding the potential value of SEA to development effectiveness.
- Lack of institutional experience of using systematic decision-making tools such as SEA.

These two challenges can be significantly addressed by capacity development for SEA. While an external partner can support local efforts, it cannot substitute them. What external partners can do is to *support* the development of capacity.

Box 7.1. Basic principles of capacity development

1. Don't rush!
2. Respect the value system and foster self-esteem.
3. Scan locally and globally; reinvent locally.
4. Challenge mindsets and power differentials.
5. Think and act in terms of sustainable capacity outcomes.
6. Establish positive incentives.
7. Integrate support into national priorities, processes and systems.
8. Build on existing capacities rather than creating new ones.
9. Stay engaged under difficult circumstances.
10. Remain accountable to ultimate beneficiaries.
11. At all stages, ensure that capacity is built to both improve skills and use the outputs from these skills. Improved analysis with out the capacity to use the analysis is of little value.

Source: Adapted from Lopes (2003).

Several other important principles of effective capacity development for SEA are:

- **Development outcome (result) orientation:** The final goal of capacity development for SEA is not a "well done" SEA but "better" decisions that result in contributions to development impact, overall development goal achievement (*e.g.* poverty reduction, improvement of living conditions).

- **System orientation:** Capacity development is not restricted to skills of single individuals but addresses organisations as well as interacting systems such as societies (*e.g.* including stakeholders and NGOs and organisations).

- **Learning orientation:** Capacity development should address the capabilities to continuously improve the decision making and implementation process, and implies adopting mechanisms to learn from reality check monitoring and evaluation as well as from previous experiences.

- **Trust-building:** All participants, especially decision makers, involved in an SEA process should be able to gain confidence in the potentials and benefits of SEA-supported decisions.

7.2. Mechanisms for developing capacities for SEA in partner countries

A capacity needs assessment will identify the kind of support for SEA that might be required and appropriate in a country (*e.g.* Case example 7.1).

Case example 7.1. Capacity-building needs assessment for the UNECE SEA Protocol Implementation in five countries in the Eastern Europe, Caucasus and Central Asia region (EECCA)

Background and objective

UNDP Regional Centre for Europe and CIS, together with the Regional Environmental Centre for Central and Eastern Europe (REC) have implemented the regional project "*SEA – Promotion and Capacity-building*". This assists the signatories of the UNECE SEA Protocol to the Espoo Convention in EECCA region – Armenia, Georgia, Moldova and Ukraine (as of 6 September 2005), as well as Belarus, in their efforts to adopt the requirements, and eventually ratify, the SEA Protocol. During the first stage of the project (May-June 2004), national consultants (representatives of the respective environmental ministries and/or of NGOs) were commissioned to undertake national needs assessments to assess the status of preparation for the SEA Protocol implementation in each of the five countries.

Approach

Different stakeholders (from various ministries, academia, NGOs and research institutes) agreed the capacity development needs during the national workshops held in all project countries. The national assessments comprised:

- Identification of plans and programmes that fall under the scope of the SEA Protocol.
- Analysis of current environmental assessment provisions.
- Description of any possible future changes in this legal framework.
- Analysis of strengths and weaknesses of the current system and opportunities for future development/improvement.
- Analysis of the priority issues for effective implementation of the Protocol.
- Identification of key players in SEA reform.
- Description of past, ongoing and planned initiatives to build SEA capacity in the country.
- Analysis of the level of consultations with the environmental and health authorities and with the public during the planning and SEA process.
- Identification of the stages of the planning and environmental assessment process, in which consultations and public participation is legally required and applied in practice.

> **Case example 7.1. Capacity-building needs assessment for the UNECE SEA Protocol Implementation in five countries in the Eastern Europe, Caucasus and Central Asia region (EECCA)** *(cont.)*
>
> - Recommendations for the most effective focus of the UNDP and REC project; Selection of the most crucial capacity development activity to be implemented within an ongoing project.
>
> **Outcomes**
>
> The following capacity development needs were identified:
>
> - *Clear definition of SEA-related terms* (e.g. plan, programme and policy) – crucial for further SEA system development.
> - *Development of the legal framework for SEA* – either by developing new laws or by amending existing ones; identified a need to harmonize legislation among Caucasus countries and with the EU legislation. Several countries expressed the need for assistance in legal reforms.
> - *Development of SEA national guidance, methodologies and training materials* for different SEA process stages.
> - *Training* seminars and workshops on SEA-related issues/topics, targeted at different stakeholder groups.
> - Demonstration of SEA through pilot application projects.
> - Sharing lessons between countries on SEA application to different procedures.
> - *Creation of EIA/SEA national centres* (Armenia and Georgia) responsible of conducting seminars, training, developing educational and methodological documents, advertising campaigns, full training of specialists for the environmental assessment, licensing, networking, etc.
> - Development of accreditation system for certifying the experts eligible to perform SEA.
>
> Based on the needs assessments, the following capacity-building activities, are being implemented:
>
> - Pilot SEA of the Yerevan City Master Plan (Armenia) and of the National Tourism Development Programme (2006-10) (Belarus).
> - Development of national capacity-building manual for SEA Protocol implementation (Georgia, Moldova and Ukraine).
>
> Findings of the needs assessment process formed a base for development of capacity-building strategies for the SEA Protocol implementation, assisted by REC, UNECE Espoo Convention Secretariat and UNDP.
>
> *Source:* Dusik et al. (2004).

Support to SEA capacity-building can involve a number of activities (see Table 7.1):

- **Technical training** on SEA principles, potentials and methods. Assuming quality content and an appropriate selection of participants, this is a direct way to enhance in-country capacity for carrying out SEA. The main target groups might be individuals and organisations involved technically in planning, development, assessment or environmental management (Case example 7.2). To reach a wide audience efficiently, training-the-trainer schemes might be useful; and training tailored according to themes or application and their specific issues, functions and needs, for example for a sector

Table 7.1. **Capacity development framework for SEA**[1]

	System capacity	Institutional capacity	Human capacity
Key elements	Frameworks within which institutions and individuals operate	Ability of an organisation to effectively operate within the given system	Skills and expertise of individual persons and their motivation
Objectives	Develop overall legislative and regulatory, frameworks Improve inter-institutional co-ordination Create enabling environments for development of entire system	Develop organisational performance and functioning capabilities	Change attitudes and behaviours Develop skills Support long-term motivation and commitment
Examples of specific interventions	Legislative, policy and regulatory reforms Practical guidelines to assist interactions between key players in SEA process Effectiveness reviews of the entire system Monitoring	Institutional audits Internal management guidelines Improved working conditions (*e.g.* tools and means of communication)	Training (initial awareness raising and basic skill development) Professional development (on-job training and acquisition of advanced know-how) Professional certification Develop networks
Examples of cross-cutting interventions	Awareness raising about benefits of good practice Platforms for regular professional debate and policy dialogue between the key stakeholders (*i.e.* professional networks or regular conferences to review and discuss states of practice) Pilot projects that test proposed changes in legislation or guidance, are implemented as part of inter-institutional learning and involve local expert through on-job training Award schemes that identify and appreciate best practices		

1. Adapted from Dusik *et al.* 2004.

Case example 7.2. **SEA training course in China**

Background and objective

China has introduced SEA legislation but has limited capacity to implement it. In response, The World Bank Institute, the International Association for Impact Assessment (IAIA) and other local and international partners, combined to develop a modular SEA training course. It is aimed at government officials responsible for making and implementing PPPs at the local, regional, and national level; and professionals involved in conducting SEA. The training is designed for distance learning and has been prepared in Mandarin.

Outcomes

The course has been offered in China for a number of years to support the newly approved Environmental Assessment Law, which requires the application of SEA to plans. Through training of environmental professionals or decision makers in general administration, the course has raised knowledge of SEA and awareness of its significance. Trainers have been trained and will continue the dissemination of SEA knowledge. The course has enhanced the skills of SEA practitioners by introducing the latest international and domestic SEA experiences and practices, and fostered partnerships with, and strengthened the capacity of, relevant government agencies and training institutes.

For more details, see *www.worldbank.org/wbi/environment/sea*.

---

such as transport or energy. In a specific context, a draft capacity development manual to support the implementation of the Espoo Convention SEA Protocol on SEA has been produced by the Regional Environment Centre for Central and Eastern Europe (REC) and the Espoo Convention Secretariat, with support from the European Commission and UNDP (see *www.unece.org/env*).

- **Awareness raising workshops** training on potentials and principles of SEA (see Case example 7.3). These can sensitise decision makers (both in donor agencies and in donor-

> ## Case example 7.3. SEA development in Mozambique
>
> **Background and objective**
>
> At the request of Mozambique's Ministry for Co-ordination of Environmental Affairs (MICOA), the Netherlands Commission for Environmental Impact Assessment advised on the environmental assessment of infrastructure for the transport of bulk goods from a titanium smelter in Chibuto, north of Maputo. A SEA was recommended to provide an integral assessment of all interests to inform a decision on transport infrastructure.
>
> **Outcome**
>
> The influence of the SEA on planning is not yet clear. But it has stimulated interest within MICOA in the potential role of assessment approaches in development planning. As a result, Mozambique has become one of the pilot countries under the World Bank-Netherlands' Partnership Programme. Stakeholders (including various ministries) agreed that SEA would be the preferred approach to use in the future and that capacity-building is required. The needs assessment resulted in an action plan, including:
>
> - Identification of a number of SEA pilot applications.
> - Awareness-raising training and technical capacity-building for those involved in the pilots.
> - Evaluation of the pilots, leading to recommendations on SEA application and guidelines in Mozambique.

> ## Case example 7.4. UNDP initiative for SEA capacity-building in Iran
>
> **Background and objective**
>
> In 2004, in response to a request from Iran, UNDP agreed a project with the Iranian Department of Environment entitled "Sustainable Development Strategy and Strategic Environmental Assessment: enabling activities and capacity-building". The initiative will help devise an SEA system in Iran. It builds on an earlier joint Government and UNDP project that established EIA guidelines and procedures.
>
> **Approach**
>
> The new project focuses on:
>
> - Capacity-building through provision of financial and technical support as well as facilitating access by stakeholders to knowledge and experience on SEA for regulation setting and training purposes.
> - Helping create a national regulatory framework for undertaking SEAs.
> - Creating an enabling environment to put in place clearly defined, participatory processes for conducting SEAs.
> - Building technical capacity within government, private sector and non-governmental players to conduct SEA studies and review SEA reports.
>
> Source: www.eiairan.org/SEA/Printable/SEA.htm.

supported countries) to the benefits of well-informed decision making for achieving sustainable development. They can help technical staff engaged in SEA to distil and summarise their findings and convey outcome messages in a non-technical way so that decision makers better understand these.

- **Supporting the institutionalisation of the SEA process.** Usually, this includes approaches such as:
  - ❖ Establishing appropriate regulatory frameworks (laws, regulations, guidebooks).
  - ❖ Clarifying responsibilities (which organisation does what within a given SEA; which stakeholders are involved with which entitlements?).
  - ❖ Supplying information or knowledge dissemination (*e.g.* Web sites and help desks with access to up-to-date knowledge).
  - ❖ Institutionalising networks and dialogue bodies (*e.g.* core groups, steering committees and dialogue panels) which enable stakeholders (both powerful and weak) to introduce ideas and practices with broad social benefits, and that combine resources towards a common goal (see Case example 7.5). Experience shows that it is best for capacity development to work with existing institutions for SEA matters rather than create new ones.

- **Supporting monitoring and evaluation systems** that verify whether previous PPP decisions achieve their intended results. This includes the tracking and public dissemination of information on the outcomes of PPP implementation through regular development reports that focus on sustainability and environment (see Case example 7.6). These provide support for a continuous process to make decision makers increasingly accountable and to guide development planning in a more sustainable direction. This assumes clear benchmarks or targets, as it is impossible to determine progress otherwise.

Case example 7.5. **Assessing the potential to introduce SEA in Nepal**

Background and objective

- The project aimed to stimulate debate amongst various sector ministries about poverty-environment links in the planning process.
- An informal Policy Forum was held during which participants from the National Planning Commission (NPC), sector ministries/departments (Agriculture, Forestry, Industry, Local Development, Physical Planning, Roads and Environment) and donor observers discussed opportunities to strengthen the consideration of poverty-environment links during the 10th Plan process.

Outcomes

This assessment concluded that:

- There were several possible entry points for SEA within the existing planning process: during the selection and prioritisation of programmes and projects; during mid term evaluation of the 10th Plan; and during the preparation and evaluation of future plans.
- Various stakeholders could have a role in developing and using SEA, *e.g.* NPC, sector ministries, local government planning teams, the Ministry of Population and Environment (MoPE) and public stakeholders such as NGOs.
- In addition to SEA, it would also be useful to monitor the extent to which development plans, programmes and activities are responding to poverty-environment issues.
- Sufficient capacity (both staff numbers and competencies) would need to be fostered existing within relevant agencies to develop and apply an SEA approach.
- The development of an SEA tool kit would be required.

> ### Case example 7.5. **Assessing the potential to introduce SEA in Nepal** (cont.)
>
> - An SEA (and draft poverty-environment indicators) should then be piloted – possibly by a small number of authorities (*e.g.* Ministries of Agriculture and Forestry) and local government administrations. This would provide the basis for training and capacity building and help improve the approach and enable the experience to be disseminated to a wider audience.
>
> *Since this assessment was undertaken, the level of conflict in Nepal has worsened and it has not been possible to take these ideas forward.*
>
> Source: ERM Nepal (2002).

> ### Case example 7.6. **Results-based monitoring in the water and sanitation sector in Colombia**
>
> **Background and objective**
>
> The Colombian government passed a Presidential Decree in 2004 requiring regional autonomous corporations, responsible for regional environmental management, to prepare three-year action plans, including outcome-based commitments. These new plans replace the previous three-year plans which focused mainly on administrative commitments. Examples of outcome-based commitments include indicators for child mortality and changes in incidences of water-borne diseases. These indicators are consistent with the findings of an SEA for the water and sanitation sector conducted in Colombia, conducted in 2000 by the Ministry of Economic Development for the World Bank-financed Water Sector Reform Assistance Project.
>
> **Outcomes**
>
> This SEA identified the deterioration of water resources as the first priority, due to its significant impact on human health (*e.g.* diarrhoeal illnesses estimated to cost USD 315-400 million a year). The establishment of such an institutional monitoring and reporting mechanism is an important first step in focusing attention on development outcomes in the sector.

- **Networking for sharing experiences** allows multiple stakeholders to learn from previous SEA cases and decisions (see Case example 7.7). Particularly valuable approaches include collections of best-practice cases, reference databases, networking of practitioners, or joint coaching teams for SEA processes. In general, feedback systems are crucial for learning.
- **Supporting network exchanges.** There are opportunities to support regional, self-help processes that build south-south networks for professional exchange and development, rather than relying only on northern experts to transfer information or undertake training (see Case example 7.8). An example of such an approach was the work undertaken under the Sofia Initiative on SEA (see Case example 7.7). In another case UNDP's Regional Centre for Europe and CIS, together with the Regional Environment Centre for Central and Eastern Europe (REC) have undertaken "needs analysis" reviews and workshops for SEA capacity-building in Armenia, Belarus, Georgia, Republic of Moldova and Ukraine. These were followed by capacity-building activities (implementation of the SEA pilots and development of the national SEA capacity-building manuals).

Case example 7.7. Sofia Initiative on Strategic Environmental Assessment

Launched under the Environment for Europe process, the Sofia Initiative (1996-2003) included a component on development and use of SEA in Central and Eastern European (CEE) countries. It addressed the institutional constraints and built capacity through a process of regional self-help, and provides a model that might be adapted to wider application internationally.

Key components included:

- Pilot SEA projects to test and develop methods and procedures and to highlight good practice.
- Briefing materials to explain the rationale for SEA to planners and decision makers.
- Practical guidance to explain the key elements of effective SEA to practitioners.
- Professional networking of SEA practitioners to help share lessons and benchmark performance.
- Training to build core professional capacities to undertake SEA.

The Sofia SEA Initiative was led by Croatia and the Regional Environmental Center for Central and Eastern Europe) and systematically brought together government officials in charge of EIA/SEA reforms in the region. Participants jointly defined the specific needs of the countries involved, contributed to regional and national policy debates on the introduction of SEA, tested new SEA approaches through pilot projects and shared lessons learned through reporting to various international forums. Much of the value from these exercises came from professional exchange and mutual learning.

Case example 7.8. The SAIEA node model for EA support

Based in Namibia, SAIEA is a "not for profit" environmental trust that acts as a node for EIA best practice, serving the EA needs of the 14 countries of the SADC region. It has three core areas of work with respect to EA: research, capacity-building and guiding, and reviewing clients' EA processes. SAIEA is a most effective node and has won several international awards. It is a model that could be duplicated in many regions. However, to ensure the success of such nodes, there must be a demand for EA support services within a region so that they can be self sustaining.

SAIEA is also the southern African node under the Capacity Learning for EIA in Africa (CLEIAA) initiative – a network that links together a number of such regional centres and associations across Africa that are experimenting with SEA. CLEIAA's goal is that, by 2010, all African countries have a functional EA system in place, adapted to local needs and capacity.

In their development assistance negotiations with partner countries, donors can augment and enhance demand for such node services by developing assistance strategies and promoting interventions that create a domestic need for help and guidance on SEA. Over the last five years, many developing countries have developed world class, but limited, SEA capacity. The donor community must make every effort to take advantage of this capacity and foster its continued development. Only then will the MDG objectives be more effectively obtained.

7.3. SEA capacity development in donor organisations

In the context of SEA in development co-operation, capacity development should not be restricted to the partner country. Important strategy-related decisions are being made within the donor organisations in connection with country strategies, sector programmes, and programme oriented funding, etc. However, a lack of knowledge, procedures or monitoring systems can sometimes lead to a deficient integration of environment into the strategic decisions made by donor organisations. To develop a solid SEA capacity donor organisations can use different approaches, many of which resembles those used to build SEA-capacity in partner countries:

- *Training activities for donor institution staff* on SEA application as an approach for sustainable decision making. Technical staff and senior management need to understand why environment needs to be integrated in decision making and the added value of using SEA to achieve this.

- *SEA guidelines:* An important step in building SEA capacity is to clearly spell out for which type of the donor organisation's decision-making processes an SEA is needed, how it should be conducted and what it should include. To be successful, SEA-guidelines should consider the specific characteristics of the planning procedures used in the organisation.

- *SEA support:* Access to support is often crucial for the program officer in a donor organisation managing or conducting an SEA. A support package can consist of checklists on what issues should be considered and templates on Terms of references for contracting consultants etc. Access to advice from SEA specialists within the donor organisation or via an external help desk are other examples of SEA support. See, for example, Case example 7.9.

- *Systematic reviews and evaluations:* The establishment of a review mechanism can be an important part of the SEA capacity of a donor organisation to ensure that environment is integrated into strategic decisions in accordance with established guidelines.

- *Increased donor co-ordination and exchange of experiences on SEA*: Through increased exchange of good practice cases, guidelines, training materials etc, added value can be provided. Also, development agencies can participate in events designed to promote exchange of experience. (See Case example 7.10).

7.4. SEA as a foundation for capacity development and learning societies

Capacity development involves using and increasing existing capacity. A first step is to identify and make full use of national expertise, consultancy and research capacity. It is important to anchor and integrate SEA in national institutions and systems and strengthen these – and avoid the temptation to by-pass them and seek to establish new or separate mechanisms for SEA delivery. Experience shows that this ultimately leads to a more sustainable outcome. Weak domestic capacity should not be seen as an argument for withdrawal or for pursuing an external agenda. Technical assistance needs to seek and to support innovative ways to promote in country leadership in capacity development for SEA.

SEA is an integrating approach. In applying SEA, links should be sought with other related approaches to assessment and impact analysis. Whilst capacity development is required for SEA application, at the same time, SEA itself offers potential to develop capacity for making complex development/public investment choices in relation to PPPs and major investment decisions. This, in turn, will build capacity to address environmental

> ### Case example 7.9. **Sida SEA Helpdesk – University of Gothenburg**
>
> **Issue**
>
> To increase its SEA-capacity the Swedish International Development Co-operation Agency, Sida, uses an external SEA-Helpdesk located at Gothenburg University. Its main role is:
>
> - To assist Sida with Strategic Environmental Assessments in different Country Strategies. This is carried out through an iterative process where the Helpdesk provides Sida with policy briefs, advice and comments.
> - To give general advice on key policy documents being developed by Sida.
> - To participate at training events for Sida staff.
> - To follow the international development within the field of SEA, aiming at incorporating best practice into Sida operations.
>
> A core team of environmental economic specialists conducts the work at the Helpdesk. Depending on the specific issues, different resource persons, including PhD students from partner countries, are involved.
>
> Sida also uses an external EIA Helpdesk located at the Swedish Agricultural University. Its main task is to review environmental assessments of project- and sector-level interventions supported by Sida.
>
> **Outcome**
>
> The Helpdesk model gives Sida direct access to expert advice on short notice. The Helpdesks complement and strengthens Sida's environmental capacity.
>
> *For more information see: www.handels.gu.se/seahelpdesk.*

> ### Case example 7.10. **Donor sharing of SEA experience**
>
> Each year, the International Association for Impact Assessment (IAIA) hosts an international meeting, attended by hundreds of practitioners, with training events on SEA in which agencies participate and share their experiences (*www.iaia.org*).
>
> The DAC SEA Task Team itself has provided a platform to marshal and share experience through meetings, email networking and a dedicated website (*www.seataskteam.net*) providing information, guidance and case materials.
>
> The World Bank has established a *Structured Learning Program* (SLP) on SEA. This focuses on the application of SEA approaches to World Bank and client operations, on the relationship with other World Bank instruments, and on ways in which use of SEA can add value to the outcomes. This includes a Web site (*www.worldbank.org/sea*) that emphasizes these aspects while providing links to some other material on the broader use and SEA definition.

sustainability in particular and the MDG agenda more broadly. Therefore, by helping countries to undertake SEA, donors can contribute significantly to the development of broader country capacity. SEA supports good governance, for example by contributing to more accountable decisions, increased public engagement and more transparency. In other words, SEA is "not an end in itself". It can also be a catalyst to a system of effective governance.

It is seldom necessary or desirable to deliver a technically perfect approach to SEA. In many situations, "windows of opportunity" arise where quick and simple support can contribute to the outcome of SEA. But flexibility is required to apply the most effective support, for example:

- During an ongoing PRSP development process, a need may arise for some more in depth reflection of environmental concerns. This could be achieved through dialogue and expert inputs on some alternative options which are more environmentally sound.
- Or, it might be necessary to introduce elements of sustainable development into a rapid planning process for a post-disaster reconstruction plan without unnecessarily prolonging the process. Here, it might be appropriate to concentrate on an assessment of the cumulative impacts of the range of individual reconstruction projects.

In all cases, balance is required between process and product, preferably using mechanisms that are already well established in a country, *e.g.* in connection with project related EIA. Consequently, capacity development for SEA should stress flexibility in using process elements and tools most appropriate to the existing challenges.

SEA supports reflexive decision making and PPP implementation when earlier assessments and decisions are evaluated and monitored to check that objectives and goals are sought and achieved. This, in turn, should contribute to better decision making in the future (see Chapter iii). SEA can therefore promote continuous improvement towards sustainable development. But it is a long-term process that requires commitment and a willingness to learn from previous successes and failures.

Further information and resources for capacity-building, and links to other useful websites, can be found at *www.seataskteam.net*.

ISBN 92-64-02657-6
Applying Strategic Environmental Assessment
Good Practice Guidance for Development Co-operation
© OECD 2006

References and Bibliography

Abaza, H., R. Bisset and S. Sadler (2004), *Environmental Impact Assessment and Strategic Environmental Assessment: Towards an Integrated Approach*, Economics and Trade Branch, UNEP, Geneva.

ADB (2003), *Environmental Assessment Guidelines*, Asian Development Bank (available at www.adb.org/Environment/policy.asp#guidelines).

Arif, S. (2005), *The Energy Environment Review (EER) in the Islamic Republic of Iran and in the Arab Republic of Egypt*, Presentation on Strategic Environmental Assessment in the Energy Sector, World Bank, Washington DC, March 2005.

Calow, P. (1998), *Handbook of Environmental Risk Assessment and Management*, Oxford, UK, Blackwell Science.

CEAA (1999), *Cumulative Effects Assessment Practitioners Guide*, prepared for the Canadian Environmental Assessment Agency by The Cumulative Effects Assessment Working Group, February 1999, Ottawa.

CIDA (2003), *Strategic Environmental Assessment (SEA) Handbook (Draft)*, Canadian International Development Agency, Hull, Quebec.

Countryside Agency, Environment Agency, English Nature and English Heritage (2002), *Quality of Life Capital*, www.qualityoflifecapital.org.uk.

Croal, P. (2003), *Quantitative Analysis of Poverty Reduction Strategy Papers (PRSPs) for Poverty/Environment Linkages and Integration*, Draft discussion methodology, Southern African Institute for Environmental Assessment, Windhoek, Namibia.

Dalal-Clayton, D.B. (2004), *The MDGs and Sustainable Development: The Need for a Strategic Approach*, pp. 73-90, in Satterthwaite D. (ed.), The Millennium Development Goals and Local Processes: Hitting the Target or Missing the Point? International Institute for Environment and Development, London.

Dalal-Clayton, D.B. and B. Sadler (2005), *Strategic Environmental Assessment: A Sourcebook and Reference Guide to International Experience*, OECD, UNEP and IIED in association with Earthscan Publications.

DFID (2001), *Keysheet on Mainstreaming Environment in Uganda's PRSP*, DFID, London.

DFID (2002), *Integrating Environment in Country PRS Processes: DFID Experience*, internal memo.

DFID (2004), *Contribution on the Environment and Natural Resources to Pro-Poor Growth: A Checklist Examining these Issues within a Poverty Reduction Strategy*, Department for International Development, London, October 2004.

DFID/EC/UNDP/World Bank (2002), Linking *Poverty Reduction and Environmental Management Policy Challenges and Opportunities*, DFID, European Commission, UNDP and World Bank (available at www-wds.worldbank.org/servlet/WDS_IBank_Servlet?pcont=details&eid=000094946_02091704130739).

DFID/ERM (2005), The *Kenya Education Sector Support Programme, Final Report* (Vol. 1), 19 September 2005, Department for International Development, London.

Dusik, J., A. Jurkeviciute and H. Martonakova (2004), *Regional Overview of the Capacity-building Needs Assessment for the UNECE SEA Protocol*, Project report, UNDP and Regional Environment Centre for Central and Eastern Europe, Szentendere, Hungary.

EIR (2004), *Striking a Better Balance – The World Bank Group and Extractive Industries, Extractive Industries Review*, http://web.worldbank.org/WBSITE/EXTERNAL/TOPICS/EXTOGMC/0,,contentMDK:20306686~menuPK:336936~pagePK:148956~piPK:216618~theSitePK:336930,00.html.

ERM (2002), *Linkages Between Environmental Stress and Conflict*, CSDG Occasional Papers 2; Kings College, University of London.

ERM Nepal (2002), *Linking Environment to Poverty Planning in Nepal*, Report on the Government of Nepal, Environmental Resources Management, Kathmandu.

European Commission (1999), *Guidelines for the Assessment of Indirect and Cumulative Impacts as well as Impact Interactions*, Report prepared by Hyder for the European Commission DG XI, Brussels, May 1999.

European Environment Agency (1998), *Spatial and Ecological Assessment of the TEN: Demonstration of Indicators and GIS methods*, Environmental Issues Series No. 11, Copenhagen, http://reports.eeu.eu.int/GH-15-09-318-EN-C/en/seaoft.eb.pdf.

Fernagut, M. (2005), SEA *Guidance for the Evaluation of Strategy Papers in Development Co-operation*, Prepared for the Belgian Directorate of Development Co-operation by Human Ecology Department, Free University of Brussels.

Hamilton, K and M. Mani (2004), Toolkit *for Analyzing Environmental and Natural Resource Aspects of Development Policy Lending*, Preliminary Draft, 29 September 2004, Environment Department, World Bank, Washington DC.

Hoffman, M. (undated), *Peace and Conflict Assessment Methodology*, Berghof Research Centre for Constructive Management, Berlin (www.berghof-handbook.net/articles/hoffman_handbook.pdf).

IAIA (2002), Strategic *Environmental Assessment: Performance Criteria*, Special Publication Series No. 1, International Association for Impact Assessment (www.iaia.org/publications).

IIED (2004), *Development Goals and Local Processes: Hitting the Target or Missing the Point*, International Institute for Environment and Development, London.

Kjørven, O. and H. Lindhjem (2002), *Strategic Environmental Assessment in World Bank Operations: Experience to Date – Future Potential*, Environment Strategy Papers, No. 4, World Bank, Washington DC.

Lopes, C. (2003), *Turning Dilemmas into Opportunities*, paper presented at a workshop on *The Challenges of Capacity Development in Africa*, organised by the Southern African Regional Poverty Network and UNDP, 2 April 2003, Pretoria (www.sarpn.org.za).

Lopes, C. and T. Theisohn (2004), *Ownership, Leadership and Transformation: Can we Do Better for Capacity Development?*, Earthscan Publications, in association with the United Nation Development Programme.

Naim, P. (1997a), Karachi's Electricity Plan: Need for a Strategic Assessment, in A.P. Adhikari and R.B. Khadka (eds.), (1998) (cited separately, Naim 1997a).

Naim, P. (1997b), *Thermal Power Generation Policy: A Strategic Analysis*, lecture at the National Institute of Public Administration, Karachi, 29 November 1997, IUCN-NIPA, IUCN Pakistan Office, Karachi.

NBI (2001), *Nile River Basin – Transboundary Environmental Assessment*, Nile Basin Initiative: Shared Vision Programme, May 2001; Nile Basin Initiative, Global Environmental Facility, UNDP and World Bank, World Bank, Washington DC.

Nelson, P.J. (2003), *Building Capacity in SEA in Sub-Saharan Africa*, paper presented at the 23rd annual Meeting of the International Association for Impact Assessment (IAIA'03), Marrakech, Morocco.

Netherlands CEIA (2003): *Annual Report 2002*, Commission for Environmental Impact Assessment, The Hague.

OECD (Organisation for Economic Co-operation and Development) (1997), *Capacity Development in Environment, Principles in Practice*, OECD, Paris.

OECD (2005), *Environmental Fiscal Reform for Poverty Reduction*, DAC Guidelines and Reference Series, OECD, Paris.

OECD (2001), *Strategies for Sustainable Development: Guidance for Development Co-operation*, The DAC Guidelines, Development Co-operation Committee, OECD, Paris, available at www.oecd.org/dac/guidelines.

OECD (2005), "Harmonising ex ante Poverty Impact Assessment", *Promoting Pro-Poor Growth: Policy Guidance for Donors*, OECD, Paris.

OECD/UNDP (2002), *Sustainable Development Strategies: A Resource Book*, Organisation for Economic Co-operation and Development, Paris, and United Nations Development Programme, New York, in association with Earthscan Publications, London (www.nssd.net/pdf/gsuse.pdf).

Partidario, M.R. (undated), *Strategic Environmental Assessment (SEA): Current Practices, Future Demands and Capacity-building Needs*, IAIA training course manual (www.iaia.org).

Pretty, J.N. et al. (1995), *A Trainer's Guide for Participatory Learning and Action*, IIED, London.

Rasso, T. (2002), *A Case Study of the SEA of the Single Programming Document for Estonia*, MSc thesis, Dept. of Environmental Science and Policy, Central European University, Hungary.

REC/UNDP (2003), *Benefits of SEA*, Briefing paper the Regional Environment Centre for Central and Eastern Europe and for UNDP, Szentendre, Hungary, May 2003.

Sadler, B. and R. Verheem (1996), *Strategic Environmental Assessment 53: Status, Challenges and Future Directions*, Ministry of Housing, Spatial Planning and the Environment, The Netherlands, and the International Study of Effectiveness of Environmental Assessment.

Sadler, B. (2001), *A Framework Approach to Strategic Environmental Assessment: Aims, Principles and Elements of Good Practice*, in Dusik J. (ed.) *Proceedings of the International Workshop on Public Participation and Health Aspects in Strategic Environmental Assessment*, Regional Environmental Center for Central and Eastern Europe, Szentendere, Hungary.

Sadler, B. (ed.) *Recent Progress with Strategic Environmental Assessment at the Policy Level*, Czech Ministry of the Environment; Netherlands Ministry of Housing, Spatial Planning and the Environment; Regional Environmental Center for Central and Eastern Europe (REC); and UNECE (*www.iaia.org*).

Sadler, B. and M. McCabe (eds.) (2002), *Environmental Impact Assessment: Training Resource Manual*, Economics and Trade Branch, UNEP, Geneva.

Shell International (2000), *People and Connection: Global Scenarios to 2020*, Public Summary (*www.shell.com/scenarios*).

Sida (2002a), *The Country Strategies: Guidelines for Strategic Environmental and Sustainability Analysis*, Swedish International Development Agency, Stockholm (*www.sida.se/publications*).

Sida (2002b), *Sector Programmes: Guidelines for the Dialogue on Strategic Environmental Assessment (SEA)*, Swedish International Development Agency, Stockholm (*www.sida.se/publications*).

UNDP/UNEP/IIED/IUCN/WRI (2005), *Sustaining the Environment to Fight Poverty and Achieve the MDGs: The Economic Case and Priorities for Action*, Message document to the 2005 World Summit prepared on behalf of the poverty Environment Partnership by UNDP, UNEP, IIED, IUCN and WRI.

UNEP (2001), *Reference Manual for the Integrated Assessment of Trade-Related Policies*, United Nations Environment Programme, Geneva.

Van Straaten, D. (1999), Vulnerability Maps as a Tool for SEA and Infrastructure Planning.

Verheem, R., R. Post, J. Switzer and B. Klem (2005), *Strategic Environmental Assessments: Capacity-building in Conflict-Affected Countries*, Social Development Paper, World Bank, Washington DC.

Wackernagel, M. and W. Rees (1996), *Our Ecological Footprint: Reducing Human Impact on the Earth*, New Society Publishers, Gabriola Island BC, Canada.

World Bank (1996), Regional Environmental Assessment, *Environmental Assessment Sourcebook Update*, No. 15, Environment Division, World Bank, Washington DC.

World Bank (1999), *Sectoral Environmental Assessment, Indonesia Water Sector Adjustment Loan*, Report No. E26, Rural Development Unit, East Asia and Pacific Region, World Bank, Washington DC.

World Bank (2000), *Environmental Assessment for Sector Adjustment Loan: The Case of the Indonesia Water Resources Sector Adjustment Loan*, Environmental and Social Safeguard Note, World Bank, Washington DC.

World Bank (2003), *A User's Guide to Poverty and Social Impact Assessment*, Poverty Reduction Group and Social Development Department, World Bank, Washington DC (available at: *www.worldbank.org/poverty*).

World Bank (2004a), *Strategic Environmental Assessment: Concept and Practice: A World Bank Perspective*, draft, The World Bank, Washington DC.

World Bank (2004b), *Environment and Natural Resources Aspects of Development Policy Lending*, Good Practice Note for Development Policy Lending, No. 4, World Bank, Washington DC.

World Bank (2005), *Program Document for Mexico Second Programmatic Environment Development Policy Loan*, World Bank, Washington DC.

World Bank (in press), *Integrating Environmental Considerations in Policy Formulation – Lessons from Policy-Based SEA Experience*, Report No. 32783, World Bank, Washington DC.

World Bank and Norplan (2004), Lao PDR Hydropower Strategic Impact Assessment, Final Report, August 2004, World Bank, Washington DC.

World Commission on Dams (2000), *Dams and Development: A New Framework for Decision Making*, Earthscan Publications, London.

WRI/UNDP/UNEP/World Bank (2005), *The Wealth of the Poor, Managing Ecosystems to Fight Poverty*, World Resources Institute, Washington DC (*www.iied.org* and *www.seataskteam.net*).

ISBN 92-64-02657-6
Applying Strategic Environmental Assessment
Good Practice Guidance for Development Co-operation
© OECD 2006

ANNEX A

Glossary of Terms

Baseline data: Data that describe issues and conditions at the inception of the SEA. Serves as the starting point for measuring impacts, performance, etc., and is an important reference for evaluations.

Benchmark: A standard, or point of reference, against which things can be compared, assessed, measured or judged. Benchmarking is the process of comparing performance against that of others in an effort to identify areas of improvement.

Capacity assessment: A structured and analytical process whereby the various dimensions of capacity are assessed within a broader context of systems, as well as evaluated for specific entities and individuals within these systems.

Capacity development: The process by which individuals, groups and organisations, institutions and countries develop, enhance and organise their systems, resources and knowledge; all reflected in their abilities, individually and collectively, to perform functions, solve problems and achieve objectives.

Civil society organisations: The multitude of associations around which society voluntarily organizes itself and which represent a wide range of interests and ties. These can include community-based organisations, indigenous peoples' organisations and non-government organisations.

Country assistance strategies/plans: A generic term for documents setting out the planned programme of assistance provided by a donor to a country, usually for a set period (often 3-4 years). They address how to achieve the MDGs. Produced usually in consultation with governments, business, civil society and others within the country.

Country environmental analysis: Policy level analysis of priorities, policy options and implementation capacity.

Cumulative effects/impacts: Incremental impact of an action when added to other past, present or reasonably foreseeable actions regardless of what agency or person undertakes such actions. Cumulative impact can result from individually minor but collectively significant actions taking place over a period of time.

Decision makers: Policy-making, planning and decision-making systems vary and the meaning depends greatly on national or agency circumstances and procedures. In SEA within donor agencies, a decision maker may be *i)* the Head of bilateral assistance in HQ; *ii)* the country manager/director and *iii)* the sectoral team leader in the agency with overall responsibility for delivering the product deriving from use of the instrument in Box 4.1; or *iv)* development co-operation advisors in embassies, etc. In an SEA applied by partner

145

countries a decision maker may be *i)* an official responsible for broad-scale or sectoral development plans or *ii)* an elected Councillor or Minister.

Development Assistance Committee (DAC): See text at front of guidance.

Development Policy Lending: A World Bank instrument focussing on issues such as governance, public sector management and reform of social sectors – such as health and education (see WB OP 8.60 and structural adjustment).

Direct Budget Support (DBS): Development agencies increasingly provide financial support to macro-level policies and to government budgets to assist the recipient through a programme of policy and institutional reform and implementation that promote growth and achieve sustainable reductions in poverty. The support may include a mix of general budget support and policy and institutional actions (including economy-wide reforms such as tax reforms, privatisation, decentralisation and trade liberalisation). **Direct Budget Support Agreements** are the formal DBS instruments negotiated between the development agency and recipient government.

Environment: Mostly used in an ecological sense to cover natural resources and the relationships between them. But, social aspects (including human health) are also often considered part of the environment. Issues relating to aesthetic properties as well as cultural and historical heritage (often in built environment) are frequently included. The DAC's *Good Practices for Environmental Impact Assessment of Development Projects* says EIA *should address all the expected effects on human health, the natural environment and property as well as socials effects, particularly gender specific and special group needs, resettlements and impacts on indigenous people resulting from environmental changes.*

Environmental Assessment (EA): The umbrella term for the process of examining the environmental risks and benefits of proposals. Interpretations of the scope of EA also vary, particularly regarding the social dimension. It is usual to consider the physical/biological impacts of development on directly affected groups (*e.g.* impacts on downstream water supply, displacement, and local communities or vulnerable groups). But many institutions routinely include consideration of social impacts that are mediated by the environment (such as the human impacts of water pollution). Some agencies undertake "environmental and social assessments" or separate "social assessments" to identify adverse social impacts and promote other social goals, such as social inclusion or poverty reduction. The relative importance of the different dimensions varies depending on the issue involved. In the case of a dam it is increasingly routine in EA to consider both physical/ecological and social impacts.

Environmental Impact Assessment: A process, applied mainly at project level, to improve decision making and to ensure that development options under consideration are environmental and socially sound and sustainable. EIA identifies, predicts and evaluates foreseeable impacts, both beneficial and adverse, of public and private development activities, alternatives and mitigating measures, and aims to eliminate or minimise negative impacts and optimise positive impacts. A subset of tools has emerged from EIA, including social impact assessment, cumulative effects assessment, environmental health impact assessment, risk assessment, biodiversity impact assessment and SEA.

***Ex post* assessment:** An evaluation of the results after implementation of a PPP. This in comparison to *ex ante* assessment where the results are assessed that a plan, programme or policy is expected or intended to have, *i.e.* based on prediction and extrapolation; it is a way of assessing whether a proposed project is feasible and leaves the opportunity to consider alternatives and adjust the plan, programme or policy to avoid or enhance the results.

Fragile States: Those failing to provide basic services to poor people because they are unwilling or unable to do so. Tackling poverty in these countries is vital to making the lives of millions of people better.

Good governance: Governance is the exercise of political, economic and administrative authority necessary to manage a nation's affairs. Good governance is characterised by participation, transparency, accountability, rule of law, effectiveness, equity, etc.

Harmonisation: Of aid procedures aims to reduce unnecessary burden on recipient countries and enhancement of development effectiveness and efficiency of aid by reduction of transaction cost of aid procedures among donors and recipient countries. Many bilateral and multilateral donors have international discussions about harmonisation of aid procedures and are engaged in harmonisation work (see also Section 1.4.).

Indicator: A signal that reveals progress (or lack thereof) towards objectives, and provides a means of measuring what actually happens against what has been planned in terms of quantity, quality and timeliness.

Mainstreaming/Up-streaming: Integrating environment into development planning processes.

Millennium Development Goals: Eight international development goals for 2015, adopted by the international community (*UN Millennium Declaration*, September 2000). The IMF, World Bank and OECD have endorsed the MDGs (see Box 1.1).

National Sustainable Development Strategies (NSDS): Called for in Agenda 21 and the Implementation Plan of the 2000 World Summit on Sustainable Development. The DAC defines NSDS as "a co-ordinated set of participatory and continuously improving processes of analysis, debate, capacity-strengthening, planning and investment which integrates the economic, social and environmental objectives of society, seeking trade-offs where this is not possible". Implementing an NSDS would most likely consist of using promising, existing processes (*e.g.* PRSP) as entry points, and strengthening them in terms of key NSDS principles in the DAC policy guidance. See OECD/UNDP (2002).

National ownership: The effective exercise of a government's authority over development policies and activities, including those that rely – entirely or partially – on external resources. For governments, this means articulating the national development agenda and establishing authoritative policies and strategies. For donors, it means aligning their programmes on government policies and building on government systems and processes to manage and coordinate aid rather than creating parallel systems to meet donor requirements.

Policies, Plans and Programmes (PPP): Have different meanings in different countries according to the political and institutional context. Here these terms are used generically. **Policies** are broad statements of intent that reflect and focus the political agenda of a government and initiate a decision cycle. They are given substance and effect in **plans** and **programmes** (schemes or sets of usually linked actions designed to achieve a purpose). This involves identifying options to achieve policy objectives and setting out how, when and where specific actions will be conducted.

Policy Reform: A process in which changes are made to the formal "rules of the game" – including laws, regulations and institutions – to address a problem or achieve a goal such as economic growth, environmental protection or poverty alleviation. Usually involves a complex political process, particularly when it is perceived that the reform redistributes economic, political, or social power.

Poverty Reductions Strategies/Papers: Prepared by a country government with the World Bank, International Monetary Fund and civil society and development partners. Describe the country's macroeconomic, structural and social policies and programmes over a three-year or longer horizon to promote broad-based growth and reduce poverty, as well as associated external financing needs and sources (*www.imf.org/external/np/prsp/prsp.asp*).

Poverty Reduction Budget Support: see Direct Budget Support.

Sectoral strategy: A policy framework, for the long- and/ or medium-term, which has been adopted by a government as a plan of action for a particular area of the economy or society.

Sector Wide Approach (SWAps) (Or Sector investment programmes): All significant donor funding support a single, comprehensive sector policy and independent programme, consistent with a sound macro-economic framework, under government leadership. Donor support for a SWAp can take any form – project aid, technical assistance or budget support – although there should be a commitment to progressive reliance on government procedures to disburse and account for all funds as these procedures are strengthened.

Stakeholder: Those who may be interested in, potentially affected by, or influence the implementation of a PPP. In the context of an SEA applied to development co-operation, stakeholders may include: *i*) internal staff (environment and non-environment) in donor agency and other departments in the donor country, *ii*) the partner country government, *iii*) other donor agencies, *iv*) NGOs, and *v*) civil society.

Strategic conflict assessment: A process to systematically assess the risk of conflict from different factors and to shape development in post-conflict states.

Structural Adjustment Programmes: A World Bank instrument prevalent in the 1980s that focussed on correcting the major macroeconomic distortions hindering development. Replaced by **Development Policy Lending** in 2004.

Tiering: Addressing issues and impacts at appropriate decision-making levels (*e.g.* from the policy to project levels).

ANNEX B

Assessment Approaches Complementary to Strategic Environmental Assessment

Country environmental analysis: See Chapter 2 and Box 2.5.

Cumulative impact/effects assessment (CIA/CEA): A technique designed to assess the combined effects of multiple activities, rather than the effects of specific development activities. A practitioner's guide is available from the Canadian Environmental Assessment Agency (CEAA 1999).

Energy and Environment Review (EER): World Bank tool involving analytical work on environmental issues related to the energy sector. It has been supported by the Energy Sector Management Assistance Programme (ESMAP), but as part of the Bank's country and sector assistance programmes. Three general types of EER have been undertaken:

- Full-scale – looking comprehensively at energy and environment issues in one of more sectors in a country.
- Rapid assessments, which are carried out to quickly prioritise key energy-environment issues in a country-based on existing data. This may lead to a full-scale EER.
- Targeted issues, such as fuel quality, sulphur emissions, or indoor air quality.

Full-scale EERs are underway or have been completed in Bulgaria, Egypt, Iran, Macedonia, Sri Lanka and Turkey, while rapid and more targeted EERs are underway or completed in Bangladesh, Bolivia, China, Mongolia, Thailand, Vietnam, Eastern Europe and Central Asia Region, and Latin America and the Caribbean Region. The Bank is reviewing the EER results and impacts, and to what extent, and in what ways, they can be a useful tool for influencing energy and environment policies and programmes in client countries. More information at *www.esmap.org*.

Gender impact assessment: Examines effects that PPPs have on women and men, particularly those that affect work-life balances, see: *www.womenandequalityunit.gov.uk/ equality/gender_impact_assessment.pdf*.

Health impact assessment (HIA): Determines how a proposal will affect people's health and provides recommendations to "increase the positive" and "decrease the negative" aspects of the proposal to inform decision makers (see *www.hiagateway.org.uk* for further information; for key references, see *www.iaia.org* – then click on resources, then HIA). Environmental health impact assessment (EHIA) is a comprehensive and rigorous approach to identify, predict and appraise those environmental factors which might affect human health.

Integrated Assessment: See trade-related assessments.

Peace and conflict impact assessment (PCIA): In 2000, the Conflict Prevention and Post-Conflict Reconstruction Network (CPRN) (see *www.bellanet.org/pcia*) agreed to explore the formation of a network of practitioners working in the area of Peace and Conflict Impact Assessment (PCIA). The PCIA Unit of the Peacebuilding and Reconstruction Programme Initiative of the International Development Research Center (IDRC) was given a mandate to assume a leadership role (*www.idrc.ca./peace*). For PCIA handbook, see Hoffman (undated – *www.berghof-handbook.net*).

Poverty and social impact assessment (PSIA) helps to:

- Analyse the link between policy reforms and their distributional impacts.
- Identify stakeholders and open up debate on policy.
- Identify alternative reforms to address the issues of concern.
- Consider trade-offs among reforms on the basis of their distributional impacts.
- Enhance the positive impacts of reforms and minimize their adverse impacts.
- Design mitigating measures and risk management systems.
- Assess policy reform and implementation risks.
- Build country ownership and capacity for analysis.
- Helps to facilitate transparency and accountability in analysis and decision making.

The *PSIA User's Guide* (World Bank, 2003) lists ten elements of good PSIA (see *www.seataskteam.net*).

1. Ask the right questions related to issues of significance to policy choice and impacting poverty.
2. Identify the stakeholders influenced by the policy or who influence the policy implementation.
3. Understand the transmission channels through which the policy changes have their impact (employment, prices, access to goods and services, assets, and transfers and taxes).
4. Assess the institutions determining the framework in which policy reforms take place.
5. Gather available data and information and identify additional needs.
6. Analyse impacts using organised research to identify links between objectives, policies and impacts focussing on winners and losers.
7. Contemplate policy enhancements and compensation, even abandoning the policy if the benefits are insufficient.
8. Assessing risks associated with underlying assumptions, including challenges to implementation.
9. Monitoring and evaluating impacts.
10. Fostering policy debate and feeding back into policy choice.

Good practice notes from the World Bank, DFID, and GTZ, further emphasise the importance of good process.

Poverty impact assessment: The OECD's Povnet is seeking a form of *ex ante* Poverty Impact Assessment (PIA) to harmonise donor approaches to poverty analysis and help shape poverty-focused or pro-poor policies and programmes. PIA was developed by Povnet using PSIA methodology, with inputs from the OECD development capabilities, and considering the MDGs and pro-poor growth and environmental objectives. It builds on existing methods and avoids duplication for greater harmonisation. PIA is more restricted and less participatory than PSIA, drawing on existing data and knowledge to identify gaps. If large, the need for further data collection, analysis, and participatory consultation may be identified, including a full PSIA. The PIA report (OECD, 2005) includes consideration of environmental sustainability which may show need for an SEA.

The draft PIA report (Povnet PIA Task Team) (OECD DAC, 2005) contains five sections covering:

1. Summary of the assessment and recommendations – initially these will include suggestions for filling identified information gaps and modifying the design in the final version recommendations to the decision makers.
2. Background to intervention, how it fits in with national objectives.
3. Matrix 1: transmission channels by which the intervention is implemented and impact on target populations.
4. Matrix 2: stakeholders and capabilities framework: outcomes for all relevant stakeholders.
5. Matrix 3: MDGs and other factors (pro-poor growth, improved governance, global environmental security) for overall assessment of the intervention's contribution.

Testing of the report starts in early 2006. It can be used to guide the design stage, and to inform decision makers. Agency officials should avoid applying PSIA and PIA in parallel and should ensure that concerns about the importance of environmental issues to poverty reduction and growth are integrated with poverty assessments.

Social impact assessment (SIA): A methodology to analyse, predict and quantify the impacts on human populations resulting from planned interventions. An offshoot of EIA. SIA references at *www.iaia.org* (click on resources).

Sustainability appraisal/assessment: An umbrella term that encompasses a range of equivalent terms such as sustainability impact assessment and strategic impact assessment for assessment approaches that are used to integrate or inter-relate the environmental, social and economic pillars of sustainability into decision making on proposed initiatives at all levels, from policy to projects and particularly within or against a framework of sustainability principles, indicators or strategies. For a review of international experience see *www.iied.org*.

Trade-related assessments: Sustainability impact assessment (also SIA) is a regularly-reviewed, research-based method developed for the EU and undertaken by independent external consultants, to identify the economic, social and environmental impacts of a trade agreement, and help negotiators and policy-makers integrate sustainable development concerns into trade policy. The EU launched the first SIA in 1999 in relation to WTO negotiations, and has subsequently undertaken such assessments for WTO negotiations with a range of countries. SIA reports and further information are available at *http://europa.eu.int/comm/trade*.

Integrated assessment of trade-related policies: A UNEP approach to help policy-makers and practitioners examine economic, environmental and social effects of trade policy and trade liberalisation, and the linkages between them. It identifies ways in which the negative consequences can be avoided or mitigated, and in which positive effects can be enhanced. The tool is used for exploring linkages between trade, environment and development; informing policy-makers across government departments and international negotiators; developing policy packages to integrate policy objectives on trade, the environment and development; and increasing transparency in policy-making. Both *ex ante* and *ex post* assessments can provide lessons and data for future assessment. A manual (UNEP, 2001) (available at *www.unep.ch/etu/etp/acts/manpols/rmia.htm*) presents a range of approaches including: formal modelling, qualitative analysis and other methods such as benefit-cost analysis, risk assessment, multi-criteria analysis, extended domestic resource cost analysis, life cycle analysis, global commodity chain analysis and scenario building.

ANNEX C

Analytical and Decision-making Tools for Strategic Environmental Assessment

Some examples of tools that could be used in the context of SEA approaches include:

1. Tools for predicting environmental and socio-economic effects

1.1. Carrying capacity analysis (CCA) determines the human population that can be "carried" by a particular area on given consumption levels, *i.e.* it identifies the limits to growth. The "capacity" concept is controversial with continued debate on what exactly it is, and how land can be managed to increase capacity. Ecological carrying capacity usually refers to the maximum population size of a species that an area can support without reducing its ability to support the same species in the future. More information at *www.ilea.org/leaf/richard2002.html*.

1.2. Network analysis (also called cause-effect analysis, consequence analysis, or causal chain analysis) explicitly recognises that environmental systems consist of a complex web of relationships, and that many activities' impacts occur at several stages removed from the activity itself. It aims to identify the key cause-effect links describing the causal pathway from initial action to ultimate environmental outcome. It doing so, it can also identify assumptions made in impact predictions, unintended consequences of the strategic action, and possible measures to ensure effective implementation. It is useful for identifying cumulative impacts. The technique involves, through expert judgement, drawing the direct and indirect impacts of an action as a network of boxes (activities, outcomes) and arrows (interactions) (*Source:* Therivel, 2004). For more information, see European Commission (1999).

1.3. Ecological (environmental) footprint analysis addresses the human impact on the Earth's ecosystems, measuring and visualising the resources required to sustain households, communities, regions and nations, converting the seemingly complex concepts of carrying capacity, resource use, waste disposal, etc. into an understandable and usable graphic form. An excellent handbook on this is Wackernagel and Rees (1996).

1.4. Social and economic analysis/surveys. Information on many of the key tools available for social analytical and survey work are described in the *PSIA User's Guide* for practitioners in developing countries. DFID has funded work on Tools for Institutional, Political and Social analysis of PSIA (TIPS Sourcebook) (soon to be available on the World Bank website). Most are available on the World Bank PSIA Web site: *www.worldbank.org/PSIA*.

Ministries of finance and other governmental bodies usually use general and partial equilibrium models for planning purposes. These predict how changes in the economy, due to for example fiscal reforms or exchange rate reforms, will affect demand, supply and relative prices. In general, these models can indicate changes in the use of different natural resources, such as energy use and agricultural output. In some cases, models also include effects on different forms of pollution. For more information, see http:// siteresources.worldbank.org/INTEEI/214584-1115794388939/20486164/ToolkitForAnalyzingEnvironm entalAspectsofPolicyLending.pdf.

1.5. Expert judgement of direct and indirect impacts is relatively quick and cheap, and can be used for applications including collecting data, developing alternatives from the strategic policy level to the detailed site level, analysing and ranking them, predicting impacts, and suggesting mitigation measures. One or preferably several experts with specialist knowledge covering the range of impacts of the strategic action will brainstorm/ discuss/consider the relevant issue. This is sometimes formalised, *e.g.* through the Delphi Technique which uses consecutive cycles of questionnaires of expert participants until agreement on a subject is reached (*Source:* Therivel 2004).

1.6. Geographical information system (GIS) is a tool to organise and present information. It combines a computerised cartography system that stores map data, and a database management system that stores attribute data. This allows links between the two data sets to be displayed. GISs are often only used to map data. However, they are also valuable analytical tools, *e.g.* for calculating areas and distances, identifying viewing areas from a point, constructing buffer zones around features, drawing contour lines using interpolated values between points, and superimposing maps of the above. For more information, see European Environment Agency (1998).

1.7. Land use partitioning analysis assesses the fragmentation of land into smaller parcels that might result from linear infrastructure development. It involves comparing before and after scenarios. For more information, see European Environment Agency (1998).

1.8. Mapping of transmission channels is a component of Poverty and Social Impact Assessment that identifies the channels through which a particular policy change or other major intervention is expected to affect stakeholders. There are six main transmission channels: employment, prices – production, consumption, and wages; access to goods and services; assets – physical, natural, social, human, financial; transfers and taxes; and authority. Impacts may be direct (from changes in the policy levers altered by the reform) or indirect (from reform through other channels). The nature of impacts may also vary over time, and so will net impacts on various stakeholders. More information at http:// lnweb18.worldbank.org/ESSD/sdvext.nsf/81ByDocName/Approach3Understandingtransmissionchannels.

1.9. Modelling (also called forecasting) are techniques that predict likely future environmental conditions with and without the strategic action. Modelling involves making a series of assumptions about future conditions under various scenarios, and calculating the resulting impacts. Models typically deal with quantifiable impacts: air pollution, noise, traffic, etc. Most models used in SEA have evolved from EIA techniques. Many are computerised. (*Source:* Therivel, 2004. The June 1998 issue of *Impact Assessment and Project Appraisal* (Vol. 16, No. 2) is devoted to modelling, though mainly in the context of EIA. See also European Commission (1999).

1.10. Overlay maps are obtained by superimposing maps of areas of constraint using transparencies (*e.g.* overlaying areas of importance for landscape, wildlife and groundwater protection). The overlay maps can identify areas that would be appropriate/inappropriate for development, and produce easily understandable results that can be used in public participation exercises. For more information, see European Commission (1999).

1.11. Participatory techniques for assessment are available for work with stakeholders and those likely to be directly or indirectly affected by a strategic action, so they can engage in the process of assessing impacts. They include, for example: participatory learning and action (PLA); participatory dialogues; focus groups and round tables; consensus-building, negotiations and conflict resolution. A useful guide to such techniques is Pretty *et al.* (1995). A participatory poverty assessment (PPA) collects poor people's views regarding their own analysis of poverty and the survival strategies. PPAs focus on poor people's capacity to analyse their situations and to express their priorities themselves. PPAs are an effective tool for obtaining direct feedback from the poor on a country's poverty profile and the impacts of policy reform. Guidance materials on PPA are available at *www.worldbank.org/poverty*).

1.12. Quality of life assessment (QoLA) aims to identify what matters and why in an area, so that the good and bad quality of life consequences (environmental, societal and economic) of strategic actions can be better considered. The technique involves identifying benefits/disbenefits that an area offers present and future generations, assessing:

- The importance of each, to whom, and why?
- Whether there will be enough of them;
- What (if anything) could substitute for the benefits?

The answers lead to a series of management implications from which a "shopping list" of things that any development/management of the area should achieve, and their relative importance. (*Source:* Therivel, 2004). For more information, see Countryside Agency *et al.* (2002) *www.qualityoflifecapital.org.uk*.

2. Tools for analysing and comparing options

2.1. Compatibility appraisal: Ensures that a strategic action is internally coherent and consistent with other strategic actions. This is not strictly an SEA function, more one associated with good planning. Normally two types of matrices are used:

- An **internal compatibility matrix** plots different components/statements of the strategic action on both axes, with compatibility/incompatibility between the actions marked in the cells with a tick or cross.
- An **external compatibility matrix** plots the strategic actions (as a whole) against other relevant (normally higher- and equal-level) strategic actions. Matrix cells are filled by listing those statements of the strategic action that fulfil the requirements of the other strategic actions, or explaining how the evolving strategic action should take the requirements into account. When no statements in the strategic action fulfil the other's requirements, or where they conflict, this may need to be addressed. (*Source:* Therivel, 2004.)

2.2. Cost-benefit analysis, scenario analysis and multi-criteria analysis to identify priorities and viable alternatives:

Cost-benefit analysis (CBA): A relatively simple and widely used technique for deciding whether to make a change. The technique adds up the value of the benefits of a course of action, and subtracts the costs associated with it. Costs are either one-off, or may be ongoing. Benefits are most often received over time. The effect of time is built into the analysis by calculating a payback period – the time it takes for the benefits of a change to repay its costs. In its simple form, CBA is carried out using only financial costs and financial benefits *e.g.* a simple cost/benefit analysis of a road scheme would measure the cost of building the road, and subtract this from the economic benefit of improving transport links. It would not measure either the cost of environmental damage or the benefit of quicker and easier travel to work. A more sophisticated approach to CBA is to try to put a financial value on these intangible costs and benefits. Guidance on the use of CBA is available at *www.mindtools.com/pages/article/newTED_08.htm*.

Scenario analysis/sensitivity analysis: Can be used to describe a range of future conditions. The impact of a strategic action can be forecast and compared for different scenarios – sensitivity analysis – to test the robustness of the strategic action to different possible futures. Forecasts based on current trends and/or scenarios representing trends outside the decision makers' control are generated and the strategic action's impacts are predicted based on these forecasts/scenarios. Sensitivity analysis measures the effect on predictions of changing one or more key input values about which there is uncertainty. The Stockholm Environment Institute has developed the Polestar Manual for scenarios *http://sei.se.master.com/texis/master/search/?q=scenarios&xsubmit=Search%3A&s=SS*.

Scenario planning is an example of a number of tools developed within the private sector (see *e.g.* Shell International, 2000). It is used to evaluate future, long-term, business environments and develop strategies that serve the traditional business goals of survival, maintenance and growth in competitive markets. The intention is to develop strategies that are robust enough to be able to adapt the company to shocks and surprises in the business environment. It does this through a systematic process, usually engaging external stakeholders, to consider the nature and impact of uncertain futures and important drivers/influences on changes in technological, societal, environmental, economic, political, commercial, cultural, etc., environments.

The goal of scenario planning is to assist strategic planners and policy analysts to make more resilient choices through understanding a wide range of possible futures and designing pathways to arrive at desired positions.

Key stages in this process include:

1. Agree the wide range of issues to address.
2. Identify participants (lateral thinkers).
3. Workshops and interviews of a "brain storming" nature.
4. Identify uncertainties and drivers of change.
5. Develop matrices to describe possible combinations of critical uncertainties.
6. Elaborate scenarios for each of the above combinations – again through group discussion.
7. Describe requirements (PPPs) to move towards a preferred vision and constraints to be overcome in getting there.

Multi-criteria analysis (MCA): Techniques that can assess a variety of options according to a variety of criteria that have different units (*e.g.* $, tonne, km, etc). This is a significant advantage over traditional decision-aiding methods (*e.g.* cost-benefit analysis) where all criteria need to be converted to the same unit (*e.g.* dollars only). They also have the capacity to analyse both quantitative and qualitative evaluation criteria (*e.g.*, yes/no, pluses and minuses). MCA techniques have three common components: a given set of alternatives; a set of criteria for comparing the alternatives; and a method for ranking the alternatives based on how well they satisfy the criteria. An MCA manual is available at *www.cifor.cgiar.org/acm/methods/mca.html*.

2.3. Opinion surveys to identify priorities: For methods go to *http://gsociology.icaap.org/methods/surveys.htm*

2.4. Risk analysis or assessment: Has established itself as an essential tool for the management of environmental risk. An issue for environmental risk assessment is the lack of an easily defined measure of what constitutes *harm* to the environment. In some cases definitions of environmental damage are laid down in statute, but in others appropriate criteria will need to be selected on the basis of scientific and social judgements. For a comprehensive treatment of the basic principles of environmental risk assessment and management, see Calow (1998). Many sources provide guidelines for environmental risk assessment, e.g. *www.defra.gov.uk/environment/risk/eramguide/index.htm*.

2.5. Vulnerability analysis: Assesses the impacts of a planned activity or different development scenarios on the vulnerability of an area. Vulnerability maps are produced showing degree of vulnerability for selected targets (*e.g.* people, flora and fauna, landscape). These are overlaid and "weighted" (using GIS and multi-criteria analysis) to indicate areas of high vulnerability and then related to expected levels of impact associated with different development options (*e.g.* noise increase, groundwater decline) – revealing the locations of negative impacts regarding different targets, and the alternatives with the least impacts. For further information, see van Straaten (1999).

3. Tools for ensuring full stakeholder engagement

3.1. General information, techniques, etc.: Many guidelines are available for effective community involvement and consultation, *e.g.*, *www.rtpi.org.uk/resources/publications/ConsultationGuidelines_web.pdf*; *www.iap2.org/associations/4748/files/toolbox.pdf*; *www.unece.org/env/eia/publicpart.html*.

3.2. Consensus building processes: A conflict-resolution process used mainly to settle complex, multiparty disputes. Since the 1980s, it has become widely used in the environmental and public policy arena but is useful whenever multiple parties are involved in a complex dispute or conflict. It allows them to work together to develop a mutually acceptable solution. More information is at *www.beyondintractability.org/m/consensus_building.jsp*. A short guide to consensus building is available at *http://web.mit.edu/publicdisputes/practice/cbh_ch1.html*.

3.3. Stakeholder analysis to identify those affected and involved in the PPP decision: Incorporates economics, political science, game and decision theory, and environmental sciences. Current models apply a variety of tools on both qualitative and quantitative data to understand stakeholders, their positions, influence with other groups, and their interest in a particular PPP. In addition, it provides an idea of the impact of the PPP on political and social forces, illuminates the divergent viewpoints towards proposed PPPs and the potential power struggles among groups and individuals, and helps identify potential strategies for negotiating with opposing stakeholders. Go to *http://www1.worldbank.org/publicsector/anticorrupt/PoliticalEconomy/stakeholderanalysis.htm*.

Sources of further information on SEA tools

- A modular Capacity Development Manual for the Implementation of the UNECE Protocol on Strategic Environmental Assessment is being developed by UNECE. It will be available at *www.unece.org*.

- Therivel, R. (2004), *Strategic Environmental Assessment in Action*, Earthscan: London, contains an Appendix with SEA prediction and evaluation techniques. It covers expert judgement, quality of life assessment, overlay maps, land use partitioning analysis, geographical information systems, network analysis, modelling, scenario/sensitivity analysis, cost-benefit analysis, multi-criteria analysis, life cycle analysis, vulnerability analysis, carrying capacity, ecological footprint, risk assessment, and compatibility appraisal.

- Rauschmayer, F. and N. Risse (2005), A Framework for the Selection of Participatory Approaches for SEA, *Environmental Impact Assessment Review*, 25(6): 650-666, covers: mediation, mediated modelling, consensus conference, citizens' juries and co-operative discourse.

- Finnveden, G., M. Nilsson, J. Johansson, A. Persson, A. Moberg and T. Carlsson (2005), Strategic Environmental Assessment methodologies – Applications within the Energy Sector, Environmental *Impact Assessment Review*, 23(1): 91-123. This paper covers: future studies, LCA, environmentally extended input/output analysis, risk assessment of chemicals and accidents, impact pathway approach, ecological impact assessment, multiple attribute analysis, environmental objectives, economic valuation, surveys, and valuation methods based on mass, energy and area.

ANNEX D

Selected Sources of Information on Strategic Environmental Assessment

OECD DAC Task Team Web site (*www.seataskteam.net*). This is the dedicated Web site of the OECD DAC Task Team on SEA. It gives information on working groups, resources, tools, biographies and includes provision for on-line discussions.

CIDA (*www.acdi-cida.gc.ca/ea*). Various publications on SEA and environmental assessment are available at (click on publications). These include an SEA handbook to provide guidance on implementing the federal 1999 Cabinet Directive on the Environmental Assessment of Policy, Plan and Programme Proposals (CIDA, 2003). The handbook is intended for those who may be involved in the development of a policy, plan, or programme, *i.e.* Cabinet liaison staff, environmental specialists, programme and project analysts, and policy makers.

European Union (*http://europa.eu.int/comm/environment/eia/home.htm*). This Web site provides information on environmental assessment and the European SEA Directive, policies, integration, funding, resources, news and development.

International Association for Impact Assessment (*www.iaia.org*). The site provides information on the IAIA, conferences, activities and special projects, resources, publications and reference materials (including SEA performance criteria and key citations for EA topics), and training.

Institute for Environmental Management and Assessment (IEMA) (*www.ieam.net*). The Centre for Environmental Assessment and Management at IEMA undertakes work on guidelines, training, research and projects. Web site provides access to publications on EA including the EA Yearbook.

International Institute for Environment and Development (*www.iied.org/Gov/spa*). The Web site provides downloadable papers and books on EIA, SEA and related subjects, and links to *www.nssd.net* for information on sustainable development strategies.

Netherlands Commission for Environmental Impact Assessment (NCEIA) (*www.eia.nl*). The Web site provides advisory services and related training activities to support the development of SEA in a country as well as advice on the terms of reference for SEA. It reviews the outcome, and gives coaching on SEA processes and the development of SEA systems. When applied, SEA is undertaken in the framework of the national context. The NCEIA is developing an SEA database which will provide a broad array of easily accessible information.

Regional Environment Centre for Central and Eastern Europe (REC) (*www.rec.org/REC/ programs/environmentalassessment*). The centre provides services for national SEA capacity-building and assists in implementation of pilot SEAs in countries in Central and Eastern Europe. REC facilitated elaboration of the Capacity Development Manual for the UNECE SEA Protocol and for the SEA Handbook for the EU Cohesion Policy in 2007-13.

Sida. Has published guidelines for SEA in the context of country strategies and sector programmes (available at *www.sida.se/publications*). These emphasise key links between poverty, the environment and sustainable development. See also Sida's SEA Helpdesk (Environmental Economics Unit, University of Gothenberg) (*www.handels.gu.se/seahelpdesk*), and Sida's EIA Helpdesk (Swedish EIA Centre, SLU/Uppsala) (*www-mkb.slu.se*).

Transport Research Laboratory (TRL), UK. The SEA Information Service Web site (*www.sea-info.net*), supported by the Centre for Sustainability at TRL provides a gateway to information on Strategic Environmental Assessment (SEA) and Sustainability Appraisal (SA).

UNECE. Information on EIA and SEA in the context of the Espoo Convention of Environmental Impact assessment in a Transboundary Context and its Protocol on SEA can be found at *www.unece.org/env/eia*.

UNEP (*www.unep.org*). UNEP has developed a second version of its EIA training resource manual as a focus for capacity-building. This incorporates a module on SEA (Sadler and McCabe, 2002). UNEP has also issued guidance on EIA and SEA good practice (Abaza *et al.*, 2004).

UN University (*www.onlinelearning.unu.edu*). The site provides a link to an SEA Course developed for the UN University, describing s range of SEA-tools and providing case materials and other valuable information.

World Bank (*www.worldbank.org/sea/*). This Web site provides information on SEA structured learning programme; understanding SEA; SEA guidance, general reference documents and country and sector specific documents; external SEA links; news and events; and questions and requests. In addition, (*www.worldbank.org/cea/*) provides information on country environmental assessment as one of the key country-level diagnostic tools designed to enhance the World Bank's knowledge of the environmental aspects of client countries' development and their environmental management framework, capacity, and performance.

OECD PUBLICATIONS, 2, rue André-Pascal, 75775 PARIS CEDEX 16
PRINTED IN FRANCE
(43 2006 14 1 P) ISBN 92-64-02657-6 – No. 55287 2006